Watershed Management for Drinking Water Protection

A joint publication of the Australian Water Association and the American Water Works Association

American Water Works Association

australian water association

Printed in China.

DISCLAIMER

This book is provided for informational purposes only, with the understanding that the publishers, editors, and authors are not thereby engaged in rendering engineering or other professional services. The publishers, editors, and authors make no claim as to the accuracy of the book's contents, or their applicability to any particular circumstance, and accept no liability to any person for the information provided in this book, or for the loss or damages incurred by any person as a result of reliance on its contents. The reader is urged to consult with an appropriate licensed professional before taking any action or making any interpretation that is within the realm of a licensed professional practice.

Editor: Chris Davis
Contributors: Michael Anderson, Chris Davis, Dan Deere, Ali El Kady, Pat Feehan, Robert Ford, Rhonda Harris, Rod Holme, Clairly Lance, Rachael Miller, Jerry Schulte, Ariamalar Selvakumar, Perri Standish-Lee, Joyce Perdek Walling, Rob Walton, Bruce Whitehill
Designer: Andrea Smith
Cover design: Matt Davis
AWWA Project Manager: Gay Porter De Nileon
Production Editor: Cheryl Armstrong

Publishers:

Australian Water Association
PO Box 222, St Leonards NSW 1590, Australia

American Water Works Association
6666 West Quincy Avenue, Denver, CO 80235-3098

National Library of Australia Cataloguing-in-Publication entry:

Davis, Chris, 1948-
Davis, Matt, 1977-
Smith, Andrea Gai, 1979-
ISBN 9781921335037 (pbk)

Catchments – Australia
Water supply – Australia
Water supply – North America
Water storage – Australia
Water storage – North America
Watersheds – North America
Water quality management – Australia
Water quality management – North America

Dewey number: 363.610994

US Library of Congress Cataloging in Publication Data

Watershed management for drinking water protection / Editor, Chris Davis.
p. cm.
Includes bibliographical references and index.
ISBN 978 1 921335 03 7
1. Water quality management. I. Davis, Chris.

TD365.W73 2008
363.6'1 dc22

2008027980

FOREWORD

Chris Davis

This publication, prepared jointly by members of the Australian Water Association and the American Water Works Association is, we hope, the precursor of several more collaborative works. The AWA and AWWA have a memorandum of understanding in place which provides, inter alia, for collaboration on areas of mutual concern.

Although communities in both North America and Australia generally enjoy the benefits of safe, palatable drinking water, practitioners in those areas are only too keenly aware of the potential and actual challenges to drinking water quality, especially in the watersheds (or, for Antipodeans, catchments).

In March 2005, a joint watershed management seminar was held in Honolulu, Hawaii, and the delegates thrashed out a framework for addressing watershed management from the perspective of drinking water quality protection. A commitment was made to produce a practice guide arising from the workshop material. That proved to be a greater challenge than had been anticipated, so the collection, collation and editing has taken three years.

The product has been worth the wait, however, and the material contributed has some kernels of watershed wisdom which will be invaluable to practitioners across the globe. In a few years time, we can contemplate revisiting the topic to ensure that the material is refreshed and still representing current best practice.

TABLE OF CONTENTS

LIST OF TABLES & FIGURES

LIST OF TABLES

LIST OF FIGURES

1 INTRODUCTION

Robert Ford

The provision of safe drinking water is much more difficult and complex than providing most other commodities or services to customers. Ready availability of a safe, reliable water supply affects the health of whole communities.

Other commodities can usually be tested and inspected before delivery, using statistical methods to ensure compliance with specifications. Providing safe, pure water through a water supply system is much more difficult. The raw water is obtained from watersheds that contain animals, usually contain human habitation and agriculture and, in many cases, industries and towns that generate polluted storm runoff. Shallow underground sources often receive their water from surface infiltration that is subject to the same type of contamination.

It is not possible to test each batch of water for all possible pathogenic organisms that may be present, in the way manufactured items are sampled and tested. Similarly, it is not possible to test each batch of water for all possible chemical pollutants before dispatch. Water is continuously treated and distributed to customers. Sampling programs can only collect and examine a minute fraction of the product and there is normally no ability to recall contaminated product. Even if testing for a range of possible pathogens was to be undertaken, most tests have a turnaround time of days, by which time the water has usually reached the customers and been consumed. In addition, the fact that no pathogenic organisms were detected is usually no guarantee that the batch is really free of pathogens, because of the huge volume of water in any batch compared to the volume of the sample.

Other techniques to provide assurance of the safety and purity of water must be employed. The most important is the application of the concept of multiple barriers, with monitoring and surveillance procedures designed to determine the integrity of each barrier, coupled with an understanding of the critical control points in the harvesting, storage, treatment and distribution operations.

For example, an appropriate disinfection process may be shown from laboratory studies to reduce the number of infective pathogenic organisms present by two or three log units (i.e., by 100 to 1,000 times). Instead of testing for pathogens in the treated water, which would be very difficult, given the extremely low level of active organisms likely to be present after disinfection, the monitoring procedure is used to determine that the treated water has been exposed to the appropriate level of disinfectant for the appropriate time. If this can be guaranteed then the water supplier can be satisfied that successful treatment by this part of the process has been achieved.

By providing multiple barriers

that, in total, create a greater level of pollution removal/reduction than the minimum required, the water supply operator can be confident that safe water is being provided at all times, even if one barrier fails; provided there are systems in place that monitor the performance of the barriers and report any failure immediately.

Source protection is usually an effective way to protect and maintain water quality and prevent pollution that will eventually require costly treatment and removal. In this sense the management of watersheds and aquifer intake areas that provide raw water is the first barrier in the multiple barrier chain.

Most watersheds and aquifer-intake areas are not under the complete control of the water supplier but are managed by a number of agencies and private landowners. These other stakeholders have a range of interests and objectives that are usually different from, and sometimes contrary to, the objectives and interests of the water supplier. Watershed management then involves not only managing water supplies but also proactive involvement and engagement with other stakeholders. Such activities usually take a long time, as the water supplier needs to identify the key stakeholders, create networks and develop trust and understanding with key decision makers and community leaders.

This requires dedicated, patient staff, skilled in negotiation techniques, who have a detailed knowledge of their watersheds and the potential risks, and who can offer effective trade-offs without compromising raw water quality. These staff members need to be supported by appropriate resources that include technical data, specialist experts (either in-house or specialist consultants) and water quality modelling. They also need a good up-to-date knowledge of the behaviour of pathogens in the environment. They must not only know the current hot spots in their watersheds but must also be alert to future development and potential land use changes. These are often difficult to recognise as they usually occur as small incremental changes over periods of years. Consequently, routine monitoring systems that track changes in land use and water quality over long periods of time are required.

A close and trusting relationship with the land planning agency and municipal authorities is critical so that intended changes to land zoning or initial, tentative inquiries about development in the watershed are referred to the water supplier and discussed at the earliest possible moment. Such early discussions can often avoid protracted and expensive conflict by alerting the potential developers to the risks and concerns before they have become financially committed to the development.

Effective watershed management also has the opportunity to substantially reduce future operational and capital expenditure. Water treatment facilities are expensive to build and operate. As the complexity of the treatment process increases to deal with increasing levels of pollution or increasing levels of risk, the cost of the facilities increases. If watershed management can reduce the risk of pollution or provide a raw water quality that is less variable, then the financial benefits to the supplier and, ultimately, the community, can be substantial.

This book aims to provide tools for water supply watershed managers to help target their use of resources to be more effective in reducing the risk of pathogens or chemical contaminants reaching consumers. Protection of raw water quality results in less risk to the community and lower costs for provision of safe water. Many contamination incidents, although classified as a treatment failure or inadequate disinfection, are primarily the result of contamination of the source water.

This book includes details of practices that are many in the industry suggest to be best practice. They have been assembled and reviewed by a group of practitioners from the USA, Australia and Canada. It must be recognised that no two watersheds or aquifer intake areas are identical, so these practices need to be considered and adapted to suit local conditions.

The book has been divided into

sections that describe the specific areas of interest in management of watersheds and, where appropriate, suggest possible options for practitioners to consider. No area can, however, be considered in isolation, and there is a number of themes such as risk management and communications that are common to all sections. Case studies have been included to provide concrete examples of the applications and links to web-based resources or reference provided if more information is required. Each chapter also includes an overview of what we should know, to help coalesce research priorities.

The chapters include best practice for all the key aspects of watershed managment. As well, some information is provided on emerging tools that are becoming available to assist water suppliers in the area of watershed management.

Provision of a water supply involves the management of risk. It can take many forms, and this book attempts to provide assistance in identifying and managing risk. While there are explicit sections on risk management and identification, most sections address identification and management of risk in one form or another.

It is hoped that this book will be of value to water industry practitioners and lead to greater recognition of the benefits of watershed and aquifer recharge area management.

2 FUNDING FOR WATER INFRASTRUCTURE

Rhonda Harris, Rod Holme

CONTEXT

Funding for water infrastructure from the source to the discharge has traditionally been a serious concern for users. Issues in dealing with funding fall into three broad categories and include:

Funding

- Lack of funding
- Identification of funding sources
- Competition for funding sources
- Sustainability of funding sources.

Costs

- Internal costs and appropriate allocations to meet the overall needs of the system, from source to delivery; collection; and to discharge
- External costs.

Who pays

- Who pays for the use of the asset?
- The true economic cost of the asset, from a watershed perspective.

An overarching issue that stretches across all aspects of water infrastructure is the synergy between the environment and the existing and needed programs.

This chapter attempts to address these funding issues on a global scale, with some specific best practices that may be implemented to help communities, municipalities, water utilities and water companies to deal with the challenges of providing safe drinking water and wastewater treatment to protect the environment.

Funding Issues

These specific issues deal with the money — where it comes from, how to get it and how to keep it coming in — in order to develop, operate, maintain and protect the watershed. Issues of funding are linked to costs and payments, which are discussed later in this chapter.

A prevalent theme throughout the world is the lack of available funding for water infrastructure. This theme is overarching; both developed and developing countries must generally deal with some form of lack of funding. For discussion purposes, these differences can be viewed separately.

In developed countries, the basic infrastructure is predominately in place, and the funding issue deals with operations, maintenance, rehabilitation and replacement of aging facilities and systems. As environmental and health regulations in developed countries continue to become more stringent, the cost of this maintenance and replacement increases. Key issues are determining the best sources of funding and the most efficient ways to utilize the funds received.

In developing countries, the priority is often to provide basic water supply, frequently from a well, with little regard to source protection. There is a need to extend the source-to-tap concept to water supply in developing countries to reverse the tendency for reactive and hence far more expensive source protection measures. While this comment applies particularly to small systems, it is also likely to apply to large systems in rapidly developing countries where reactive source protection will be even more expensive.

In many developing countries, forestry/deforestation is the main watershed issue, so integrating this established issue with drinking water source protection may be beneficial.

Competition for funding sources is increasingly tight — grants are less available, and customers are often unwilling to approve increases in rates or water tariffs. One reason for this is the lack of public education and understanding about the value of water and what it takes to provide safe drinking water. In some cases, systems are failing as a result of deferred maintenance. Some utilities do not have an incentive in place for the best allocation and use of their water assets. Customers do not appreciate what it takes to get the water to them — they take the asset for granted, and may not understand the importance of watershed protection as part of the overall picture of water service delivery.

Cost Issues

In trying to educate the customer about the true costs of a system, two issues surface:

- Internal costs and appropriate allocations to meet the overall needs of the system, from source to delivery to collection and to discharge
- External costs.

Internal costs include the fixed costs of a system, such as:

- Cost of the necessary assets, including land, water rights, hardscape (facilities) and transport infrastructure (pumps and piping), appurtenances and equipment
- Cost of employees to operate and maintain the assets, and to provide support services
- Source protection costs, which frequently include incentives to change or improve activities within a watershed.

External costs include the peripheral support for the system from outside sources, including:

- cost of contractors to support the system, including legal, financial and engineering services; and
- investment sources, including capital markets, grants and bonds, used as catalysts to mobilize flows of funding for a system.

Who Pays

Who pays for the use of the asset is always a source of concern. The ideal situation for any system is for the user to pay for the water used, the polluter to pay for wastewater generated, and for other users to pay their way too, e.g., recreators. A system should aim for full cost recovery if at all possible. This cost recovery should include the fixed costs as well as the variable costs of a system, as well as the opportunity cost of the resource. This means that the manager needs to consider the watershed as a business asset in determining its true economic cost for use in determining the actual full system cost.

In large systems in developed countries, it appears to be most usual for watershed protection costs to be part of a utility's regular rate structure. Since watershed protection regulation is evolving in most jurisdictions, accompanying funding is often available, either as an overall regulatory program, or in response to specific utility initiatives.

In developing countries, case studies from bodies such as the World Bank indicate that a link is required between water charges paid by downstream users and investment in watersheds upstream. Even where irrigation fees or water taxes are known and accepted, payment for watershed protection is new, with growing acceptance by larger stakeholders.

OPTIONS

While water systems are natural monopolies by their very nature, there are still some market-driven forces working on a system's ability to pay for its costs with user fees. These market forces include:

- Customers' ability to pay (which affects cash flow)
- Regulatory changes and enforcements that affect the cost of business
- Limited financial expertise
- Limited borrowing ability.

In some systems it is more difficult to achieve full cost recovery. In particular, small systems bear a higher cost for the system in relation to the numbers of customers. Typically, small systems use a greater length of pipe to serve each connection. Many of these customers, who are low income or elderly, cannot pay the level of user fee the system would have to charge to cover the cost of financing, constructing and operating the system. In addition, these small systems do not or cannot put aside funds for maintenance and replacement, and hence they have no reserve for replacement when the system deteriorates, creating an emergency situation.

Even the financing itself is more costly because small systems have smaller capital needs, and thus basic transaction costs are relatively high in proportion. In addition, commercial banks are usually unwilling to make loans to small systems, contributing to the lack of funding availability.

The true economic cost of the asset, from a watershed perspective, is important in determining the issues of cost and payment.

A system must have a firm grasp on the total picture of the asset, including:

- Status of the system characteristics, such as present and future level of system coverage
- Need for repair, renovation, replacement or upgrade
- Current sources of funding and the ability to preserve future funding
- Growth projections and needs
- Status of regulations
- Extent of future budgets
- Current and future employment targets
- Current collection rates
- System deficiencies
- Incentives and programs for watershed protection.

The suitability of different project-specific funding options will depend on the relationship between the risks, motivations and cash flow of a system. For example:

- A system requiring major systematic overhaul and large amounts of capital in the near term must have a prospect of sufficient collections and firm cash flow to induce associated large scale institutional risk taking. This prospect relates to issues such as rate of return, management autonomy, cut-off rights, etc.
- A governmental preference for continued control and public employment may be inconsistent with remedying the sources of the system difficulty. This prospect may suggest the use of more limited or long-term measures of repair of system problems.
- In particular, the factors that are a major component of both the short- and long-term shape of the system include the water demand curve; where the demand is coming from; who is providing payment in support of that demand; and the effects on the system overall.

Water systems must be developed, if at all possible, to avoid any material risk of cash short-falls. There must be future capacity or capability for rapid adjustment in capacity. Failure to have these capabilities in place can lead to capacity shortages or significant tariff increases. Some means of making capital available to systems include:

- Private activity bonds
- Credits from multiple sources, both public and private
- National guarantees, in the form of grants
- Revolving fund programs that are able to leverage the funds available for

loans with matches, issuance of bonds and interest earnings.

Another concept of reducing capital needs is to reduce infrastructure needs by using asset management, distributed and non-structural technologies and integrated water resource planning. Implementation of a comprehensive asset management strategy can save as much as 20% of the operating and capital budgets of systems by reductions in unnecessary system replacements and higher cost emergency repairs. Utilization of a wide range of distributed and non-structural technologies, including point-of-use water treatment for isolated homes or neighbourhoods, low impact development technologies, wetland restorations, water conservation measures, stormwater retention, and others, can save even more operating and capital costs.

The use of integrated water resource planning, including restructuring regulations and systems onto watersheds, where water is viewed in an integrated, holistic framework, leads to both cost savings and environmental improvements. Looking at the big picture from source to tap and back to source, for synergy between existing programs and protection of the environment, provides a sustainable future for the watershed.

BEST PRACTICE

Funding

Current best practice for funding sources covers a variety of approaches, including:

- Setting a pricing strategy based on the investment in the source and its protection
- Providing rebates or incentives for source protection, water conservation, etc.
- Setting tariffs to achieve full cost recovery
- Utilising bond sales, grants and loans to the fullest benefit to smooth out spikes in funding needs
- Adding a percentage tariff to be specifically allocated for watershed protection (ecosystem tariff)
- Working with environmental groups, such as The Nature Conservancy in the USA, to purchase watershed areas, and providing source protection by limiting development in the watershed area
- Showing a commitment to the watershed by spending money on appropriate on-ground works to provide the best use of the source water and to treat wastewater to a level that does not harm the environment upon discharge
- Providing transparency in all funds allocations
- Establishing partnerships, including support through knowledge sharing and capacity building
- Using a risk-based approach to assessing threats to drinking water in the watershed and using established partnerships to determine the most cost effective way to eliminate the threats.
- Once a system has the funds, can it sustain the funding process over an appropriate period of time? Is full cost recovery possible? Can the system generate enough funds to meet both its short term and long term needs? Is the system living on growth and neglecting the aging of its infrastructure? Sustainability is a serious issue for any system.

A key best practice in dealing with sustainability is education. The watershed manager must develop in its customer base a knowledge of the importance of protection and sustenance of the system, both from a holistic viewpoint and a specific component standpoint. Customers have to be given information to understand the true costs of a system, including:

- The cost of protecting the water source, typically including the watershed
- The cost of providing safe drinking water
- The cost of delivering the water to the customer, in adequate volume and pressure
- The cost of collecting wastewater and transporting it for treatment
- The cost of treating the wastewater,

putting clean water back into the environment and dealing with the residuals from the treatment process in a safe and efficient manner
- The cost of assessing and reducing risks in the watershed.

Costs

A key component of the success of a system that uses the best practices in addressing its true costs is transparency. If the process is transparent, then the system is more likely to be creditworthy, and lending agencies are certainly more likely to be willing to provide financial capital.

Conclusion

There are several mechanisms currently available to a water system in terms of funding for its needs. Becoming more flexible and looking beyond the traditional sources will allow the system to utilise some different funding mechanisms and address its needs in a more holistic manner. Including watershed protection costs in already established cost recovery mechanisms is an essential step.

CASE STUDIES

The following case study snapshots are based on experiences in a variety of different watershed and catchment protection situations and organizations, including:

- The World Bank
- Asian Development Bank
- Recommendations of the Inquiry into Drinking Water Contamination in Walkerton, Ontario
- Nature Conservancy
- New York City
- City of Philadelphia
- Halifax Regional Water Commission
- Canadian Council of Ministers of the Environment
- County of Oxford, Ontario (Groundwater Systems)

World Bank

www.worldbank.org
(search for Watershed Funding)
Note that the World Bank has a debate on output based aid (which appears to involve private funding up front) vs. input based (which appears to involve public funding up front). Extracts from World Bank reviews of past projects contain the following information.

Payments for Watershed Protection services, The Case of the Segara Basin in Indonesia:

- In general, there are no direct links between water charges paid by downstream users and investment in watersheds upstream
- Largely regarded as cost-centered in development
  - This understanding influences how the stakeholders behave towards watershed resources
- Irrigation Service Fee/water tax — known and accepted but concept of payment for watershed protection is new and the paradigm is accepted by larger stakeholders
- Key Constraints
  - Some downstream water users are poor
  - Little or no management of health or environmental risks
- Future Challenge
  - Advocating an integrated approach (government, forestry, agriculture, water, upstream service suppliers, water users, and third parties)
- Funding mechanisms — direct or indirect
  - Legislation needed to encourage environmental solidarity among water users to share in costs and benefits of watershed protection
- Ultimately the formalization of any market-based or other payment system for watershed protection will be primarily a tool for integrated river basin management rather than an end in itself
- Stakeholders must be brought together to negotiate rights and responsibilities over water.

Recommendations for scaling up watershed management in Indonesia are based on two decades of experience, which show the importance of

* Social organization
* Start small and build up
* Flexibility
* Favourable hydrological conditions
* Local autonomy over resources
* Markets and infrastructure

Beware the distorting power of input-based project funding

* Technocratic focus (agency)
* Focus on short term benefits (locally)
* NGOs should support but not lead
* Protecting the needs of the poorest remains a problem
* Let local communities keep the benefits
* Consider performance-based watershed support

- Draw on experience with payments for environmental services
- Eliminates input-based funding with its negative incentives
- Stimulates technical and institutional innovation
- Requires community resource ownership rights
- Retain technical and organizational assistance.

North American Development Bank

www.nadb.org

The North American Development Bank funds environmental projects certified by the Border Environment Cooperation Commission in an area 100 km north and 300 km south of the US/Mexico border. The two institutions work together with communities and project sponsors in both countries to develop and finance infrastructure necessary for a clean and healthy environment for border residents. Information is available at www.cocef.org/ingles.php.

These funds take the form of loans and grants, and require public participation. They require steering committees to work on the community impacts.

Grant funds are available through the Border Environmental Infrastructure Funds (BEIF) from the US Environmental Protection Agency (USEPA), but only if the project can prove positive environmental impacts on the US side of the border. If projects fit this scope, then the BEIF will provide funds on a one-to-one matching basis. Information is available at www.epa.gov/owm/mab/mexican/mxsumrpt.htm.

An example is the El Morillo Drain, an agricultural drain south of McAllen, Texas, that empties into the Rio Grande. The drainage is very salty, and detrimental to the river, so the proposed project is to build a pipeline all the way to the Laguna Madre on the Gulf of Mexico coastline and remove the salty water from the Rio Grande.

The agencies also have loans that are at market rate and relatively competitive. The loans are limited to 50 percent of the capital costs, but this limit is being raised to 80 percent. The rest must be provided by the community, in effect a matching program.

Mexico

Within Mexico itself, there are two federal programs: the Aguas Potable Alcantarillado Para Zonas Urbanas (APAZU) which will match investments at various rates, and the Programa de Devolucion Derechos de Agua (PRODDER). Communities pay the federal government for the water rights, since in Mexico all of the water is owned by the federal government, and some of the money is offered back to the communities with matches for infrastructure projects. Additional information is available from www.slp.gob.mx/cea/Paginas/Programas.htm and www.cna.gob.mx/eCNA/Espaniol/Organismos/Central/Programas/PRODDER_CNA.htm.

One key factor in all of the projects is that the utilities must become self-sustaining, and have a positive cash flow with reserves, debt service, operations and maintenance funds and repair and rehabilitation funds. In Mexico, communities must have State Congress approval for getting loans, which can be problematic at times, if there are two different political parties involved.

There is investment from four parties in Mexico:

- Comision Nacional de Agua
- State governments
- Counties (Municipios)
- Local utilities

The bond market is just now beginning to move in Mexico, and just beginning to establish credit ratings for utilities and municipios.

Asian Development Bank and the Network of Asian River Basin Organizations

One of the objectives of the Asian Development Bank Institute is to complement the Asian Development Bank's (ADB) existing research, advisory, lending, and technical assistance programs.

ADB is managing a portal web site on water (www.adb.org/Water/NARBO/default) that links to Network of Asian River Basin Organizations' charter, constitutional body, action plans, technical presentations, and members list. Materials can be reviewed at

- www.adbi.org/partnership.narbo
- www.adbi.org/catalog/index.php?tab=4&modid=256&breadcrumblabel=Water.

Some global case studies on general Infrastructure Finance can be found at www.adb.org/Documents/Reports/PRC-Urbanization-Strategy/vol3/sec4.pdf.

Recommendations of the Inquiry into Drinking Water Contamination in Walkerton, Ontario

These recommendations are from Part 2 of the Inquiry Report and cover financial aspects of source protection. These recommendations were followed by the Ontario government in developing the Clean Water Act (CWA), which is covered under Regulations. This information is not available online but can be purchased from Service Ontario Publications (No. 300013) (www.publications.serviceontario.ca/ecom).

Recommendation 7: "The provincial government should ensure that sufficient funds are available to complete the planning and adoption of source protection plans."

- Combination of public funds, water user fees and fees from pollutant dischargers
- Public funding on the basis of broader benefits but contrary to user pay concepts
- User pays — fees on water users
- Watershed-based source protection need not be very expensive
- Watershed planning is a "highly overlapping process."

Recommendation 14: "Once a farm has in place an individual water protection plan that is consistent with the applicable source protection plan, municipalities should not have the authority to require that farm to meet a higher standard of protection of drinking water sources than that which is laid out in the farm's water protection plans."

- Balance two competing interests protecting drinking water and viability of farms
- Farmers participate in source water protection plans which are then approved
- Municipalities should not then be able to unilaterally impose more stringent measures.

Recommendation 16: "The provincial government, through the Ministry of Agriculture, Food and Rural Affairs in collaboration with the Ministry of the Environment, should establish a system of cost-share incentives for water protection projects on farms."

- Voluntary measures not effective if farm profitability affected

- Incentives used in other jurisdictions — "society benefits from environmental protection and inexpensive food"
- Specific mention of abandoned wells — threat to farmers and groundwater sources.

The Nature Conservancy

www.nature.org and www.nature.org/initiatives/freshwater
The Nature Conservancy (TNC) is developing a Blue Water Award to recognize utilities that protect the environment. TNC also has a Sustainable Waters Program that works to integrate sustainable water management into all water managers' portfolios as the framework they rely on to protect the systems for which they are responsible. The vision of The Sustainable Waters Program is to help protect freshwater ecosystems by advancing water policies that secure adequate water flows in rivers, lakes and wetlands. TCN states that it is realizing its vision in the US by working with partners including federal water agencies, private energy companies, municipal water providers and state governments. Internationally it is building relationships with nongovernmental organizations, US Agency for International Development and the World Bank. Through a core group of scientists and fresh-water specialists, the Sustainable Waters Program promotes site-based projects across TNC state and country programs to implement ecologically sustainable water management.

City of Philadelphia

www.phillyriverinfo.org
The City of Philadelphia Water Department is a water, wastewater, and stormwater utility. It is revenue neutral and generates revenue through cost of service charges. The typical water/sewer/stormwater bill in Philadelphia is around $40/month. Grants from local, private, state, and federal agencies, including federal appropriations from Congress, support the demonstration and pilot projects. The water department was recently awarded $1.1 million from USEPA for a Target Watersheds Initiative grant for the Schuylkill River. It also gets matching funds from other water suppliers and organizations. It states that it tries to leverage Clean Water Act monies and programs to do a lot of its watershed protection work, since it feels that is where significant funding and regulatory power is located.

Halifax Regional Water Commission, Nova Scotia

www.halifax.ca/hrwc/watershedmanagement.html
The following notes outline the commission's approach to protecting its watersheds:

- Public ownership is best form of control
- Goal is to have all watershed land owned by either the government or Halifax Regional Water Commission (HRWC)
- 2 percent to 5 percent of capital budget is spent on watershed land purchases annually
- 95 percent of watershed land is now owned by HRWC or the government
- Potential for forestry revenue (based on keeping the forest young, healthy and diverse — water quality is the primary objective)
- Potential rental revenue
- Expense and liability of
  * Boundary maintenance
  * Occupiers' liability
  * Enforcement of regulations and private land rights
  * Property maintenance.

Information attributed to HRWC on the advantages, disadvantages and alternatives of watershed purchases (from the Canadian Council of Ministers of the Environment), is as follows:

Advantages

- High level of control over activities
- Land can be identified as private property through signage
- Violations can be dealt with as trespassing

- Potential for forestry and rental revenues

Disadvantages

- Purchase price can be high
- Expense and liability of boundary maintenance
- Occupier's liability
- Patrol and enforcement costs
- Property maintenance costs

Alternatives

- Controlling access points (rather than the entire area)
- Establishing partnerships with other municipalities, organisations, stakeholders and land users

New York City

www.nyc.gov/html/dep/watershed
New York City obtains State Safe Drinking Water Act (SDWA) funding and provides funding to watershed users to protect the watershed.

Canadian Council of Ministers of the Environment

www.ccme.ca
Canadian Council of Ministers of the Environment (CCME) is comprised of the environment ministers from the Canadian federal, provincial and territorial governments. CCME seeks to achieve positive environmental results, focusing on issues that are national in scope and that require collective attention by a number of governments.

CCME offers free downloadable publications at www.ccme.ca/publications/list_publications.html#link8.

Their publications include *From Source to Tap: Guidance on the Multi-barrier Approach to Safe Drinking Water,* which includes sections on investing in source water protection, socio-economic tools and decision support systems.

Investing in Source Water Protection

This section includes the statement "Source water protection measures, especially in the context of a drinking water program, are generally preventative in nature. Even during the best financial times, it is difficult to commit money to these types of programs because the public is generally much more accepting of funding tangible, visible results such as road repairs. However, protecting human health is an essential pursuit of stakeholders and participating partners in the drinking water program. The absence of illness is a positive result that can be measured, demonstrated and communicated."

Socio-economic Tools

According to *From Source to Tap,* "Socio-economic tools offer analysis of anticipated social and economic impacts associated with considered watershed/aquifer management policies. Tools in the category include:

- Cost and Return Estimator (CARE) by the US Department of Agriculture Natural Resources Conservation Service, which generates costs and returns for crop enterprises
- The Farm Level Income and Policy Simulation Model (FLIPSIM) by Texas A&M University simulates the impacts of alternative farm and natural resource policies on the survival and profitability of representative farms
- *A Guide for Cost-Effectiveness and Cost-Benefit Analysis of State and Local Water Protection Programs* shows how to use cost-effectiveness and cost-benefit analysis to evaluate groundwater programs.

"The tools are presented in a step-by-step fashion so those unfamiliar with formal economics can still use them."

Decision Support Systems

This section of *From Source to Tap* includes the following: "Watershed planning and management can be complex. Making decisions that benefit stakeholders while maintaining watershed objectives can be challenging. The decision-making process is multidisciplinary in nature and must integrate variables such as scientific, socioeconomic and political knowledge. Fortunately, decision support systems

(DSS) that make use of complex, dynamic knowledge from a number of disciplines, generally in a user-friendly graphical user interface, are available to decision makers. These systems allow users to organize information, design alternative watershed management plans and assess the consequences of these to stakeholders. It is recommended that people conducting long-term sustainable watershed planning and management consider the use of DSS."

County of Oxford, Ontario (Groundwater Systems)

www.county.oxford.on.ca/site/599/default.aspx

The County of Oxford consists of numerous small rural communities, each with individual groundwater systems. When the County took over responsibility for all County water systems, it undertook comprehensive surveys of all sources, and commenced implementation of protection.

All activities are funded from water rates and Ontario government grants designated for source planning activities. Oxford has rationalized its rate structures by adopting a uniform rate for defined groups of individual systems.

REGULATIONS

The following is a summary of funding opportunities in regulatory approaches in different jurisdictions, including:

- USA: US Environmental Protection Agency
- European Union: European Water Framework Directive
- Canada: Ontario Clean Water Act (Legislating and Regulating Drinking Water Source Protection)

USEPA

The Watershed Plan Builder Tool and Planning web site is iaspub.epa.gov/watershedplan/watershedPlanning.do?pageId=48&navId=35. This site includes funding information under iaspub.epa.gov/watershedplan/informationSource.do?pageId=53&navId=41. USEPA also has information at www.epa.gov/owow/funding.html including:

- USEPA Watershed Funding Programs and Additional Federal Funding Opportunities
- Tips on how to achieve sustainable funding
- Sustainable Funding Planning Tools and Training
- Resources for Non-profit Organizations
- Resources for State and Local Governments
- Resources for Funders

European Union Water Framework Directive

ec.europa.eu/environment/water/water-framework/info/intro_en.htm

The European Union Water Framework Directive states: "The best model for a single system of water management is management by river basin — the natural geographical and hydrological unit — instead of according to administrative or political boundaries."

One component of that approach is an economic analysis of water use within the river basin. This is to ensure that there be a rational discussion on the cost-effectiveness of the various possible measures.

The need to conserve adequate supplies of a resource for which demand is continuously increasing is also one of the drivers behind what is arguably one of the Directive's most important innovations — the introduction of pricing. Adequate water pricing acts as an incentive for the sustainable use of water resources and thus helps to achieve the environmental objectives under the Directive.

Member States are generally required to ensure that the price charged to water consumers — such as for the abstraction and distribution of fresh water and the collection and treatment of waste water — reflects the true costs. Whereas this principle has a long tradition in some countries, this is currently not the case in others, where derogations from the pricing principle are possible, e.g., "in less-favoured areas or to provide basic services at an affordable price."

Ontario Clean Water Act

www.ene.gov.on.ca/en/water/cleanwater/index.php

The Ontario Clean Water Act was enacted on October 19, 2006, when five provisions came into effect. Full implementation will occur as associated regulations come into effect.

According to the Ontario Ministry of the Environment, the legislation is designed to promote voluntary initiatives but does require mandatory action where needed. The legislation sets out a basic framework for communities to follow in developing an approach to protecting their water supplies that works for them:

- Identify and assess risks to the quality and quantity of drinking water sources and decide which risks are significant and need immediate action, which need monitoring to ensure they do not become significant, or which pose a low or negligible risk
- Develop a source protection plan that sets out how the risks will be addressed. Broad consultation will involve municipalities, conservation authorities, property owners, farmers, industry, businesses, community groups, public health officials, First Nations and the public in coming up with workable, effective solutions
- Carry out the plan through existing land use planning and regulatory requirements or approvals, or voluntary initiatives. Activities that pose a significant risk to drinking water sources may be prohibited or may require a site specific risk management plan. This plan will set out the measures that a property owner will take to ensure the activity is no longer a threat
- Stay vigilant through ongoing monitoring and reporting to measure the effectiveness of the actions taken to protect drinking water sources and ensure they are protected in the future.

The Act creates a drinking water support program (Stewardship Fund) to help farmers and small rural businesses take action to reduce threats to local drinking water sources. A panel that includes agricultural, municipal and conservation authority representatives will provide advice on how the drinking water stewardship program should be administered and allocated. The funding is enshrined in the legislation. Initially, C$7m was made available in 2007/08 for early action to protect drinking water, C$5m to support action to protect land and water surrounding municipal water wells and surface water intakes and C$2m to support local education and outreach related to source protection planning. Additional information can be found at: www.ene.gov.on.ca/envision/news/2006/091901.htm.

The Ontario government also committed approximately C$120m from 2004 to 2008 to support source protection planning. This includes funding to enable municipalities and conservation authorities to undertake technical studies to provide the baseline scientific information needed for source protection planning (although no decision has been made on implementation funding until the plans are complete and the scope of work is known). This funding is discretionary on the part of the government (i.e., it is not prescribed in the legislation).

WHAT SHOULD WE KNOW?

Table 2.1 on page 16 represents an outline of information gaps and suggested research needs that may lead to solutions.

OTHER RESOURCES

There are numerous resources available from many sources, some of which are included in the previous sections.

Table 2.1 Information gaps and research needs

Issues	Ideas
• Lack of financial resources • Need to better identify sources of funding • How can we demonstrate that funding is spent effectively?	• Increase price of water and specifically allocate a portion of water revenue to watershed protection. Include watershed protection when pursuing increases • Need beneficiaries to pay for watershed protection • Create a visible presence – include spending on above-ground works
• Include the principle of Polluter Pays in any market regulation	• Watersheds should be considered as business assets
• Synergy between environmental movement and watershed protection programs	• Emphasise benefits of watershed protection in terms of improved public health through drinking water protection • Maintain a balance between overall ecological protection in general and drinking water source protection in particular
• Funds allocated to watershed protection need to be sustainable	• Water charges and funding programs need long-term commitments, such as being enshrined in legislation and not one-time grants or programs subject to cancellation or political influence
• Internal promotion of the advantages of source water protection • Economic cost of failure to protect • Cost/benefits – can these be tied to environmental benefits as well?	• Adopt risk management approach to drive municipal or corporate investment in source protection (Use need for due diligence to gain management/governance support) • Decide on priorities – which is the driver?
• How to get into the competition for funding?	• Adopt a full-cost price structure based on a complete asset management plan that includes the watershed
• Is an "ecosystem" fund tax feasible or acceptable? (Note that the 80/20 Agriculture/ Urban water use statistic raises equity issues)	• Cooperate with and leverage other agency's incentive schemes, i.e., partner with nature and conservation group activities

Following is a summary of relevant resources.

- American Water Works Association (AWWA)
  www.awwa.org
- Australian Water Association
  www.awa.asn.au
- Awwa Research Foundation
  www.awwarf.org
- European Union Water Framework Directive
  ec.europa.eu/environment/water/water-framework/info/intro_en.htm
- International Water Association
  www.iwahq.org
- Michigan State University Virtual Watershed Program
  www.iwr.msu.edu/vu/watershed.html
- Ontario Ministry of the Environment
  www.ene.gov.on.ca/en/water/index.php
- Water Environment Federation (WEF)
  www.wef.org

3 INVOLVING STAKEHOLDERS & ACHIEVING CHANGE IN WATERSHEDS

Pat Feehan, Bob Cavill

CONTEXT

Water suppliers very rarely have the luxury of closed (protected) watersheds from which to extract raw water. As well, watersheds rarely have a single management authority. It is usual to have many stakeholders involved in watershed management, including those with direct land and water management responsibilities (e.g., farmers) and government and semi-government authorities with indirect land and water management responsibilities (e.g., USEPA and municipalities).

Watershed managers frequently wish to influence or change land management practices within their watersheds, to achieve raw water quality improvements.

In unprotected watersheds those directly responsible for land and water management decisions (landholders) have arguably the greatest capacity to influence water quality within the watershed. Many of these stakeholders have complementary interests (usually), although some conflict may exist.

Land managers in a watershed area can be quite diverse, depending on the watershed — from small allotment owners and rural residential, to extensive grazing properties, horticulture and farming corporations to government agencies and community service providers. No single issue or incentive is necessarily relevant to all parties. For instance, landholders in the watershed area may not use the water they are expected to protect and may thus have no incentive to protect its quality.

Cooperation and partnerships with landowners is a significant implementation component of watershed management. It's the opportunity for everyone to get involved to instigate on-ground change. However, it is time consuming, slow and often hard to justify in a hard-nosed cost-benefit sense. Its major benefit comes in assessing risk management and community attitudes to water and the water agencies.

OPTIONS

Watershed managers have a number of options for implementing and undertaking change processes and involving key stakeholders. These can be summarized as shown in Table 3.1.

BEST PRACTICE

There are many possible *best practices* that can be applied to a given situation. It is likely that no single best practice can be applied to a broad range of stakeholder involvement and change activities. It is impossible, and undesirable, to be prescriptive. Rather, application of a combination of application of a set of principles and lessons learned from case studies need to be adapted

Table 3.1 Stakeholder engagement options and opportunities

Options	Opportunities	Risks
• Do nothing	• Not a sensible option	• Loss of control over watershed
• Do it yourself	• May work for small utilities or small projects or for very high risk incidents	• Can be painful • Do it at your peril — can be lots of opposition from different groups and individuals
• Use someone else's current process (i.e., use existing groups already established in the watershed) or • Use someone else's model (not necessarily water) or • Adjust, develop and apply someone else's process to your own situation	• Uses existing networks, processes, learning	• Can't just pick up and use someone else's process — it will need adaptation • May need to calibrate data if it's from another source • Could end up not getting the priority desired • Different timelines
• Set it up and let someone else lead it	• Someone else to blame • Can be a cost-effective way of getting things done (especially for government) • Can create standardization • Can get experts to run the process for you	• Could be resource hungry • No champion for the process • Won't achieve outcome desired • Surrender control • Case study — Mt. Lofty Ranges Review that didn't work in South Australia
• Set it up and lead it yourself	• Retain control • Demonstrate leadership • Can build community trust and involvement • Can invest in ways to get the community on side, e.g., sponsor community events • Can save money • Case study — Wivenhoe dam upgrade (Brisbane, Queensland, Australia) • Case study — Twin tunnels (Vancouver, British Columbia, Canada)	• Can get part way through process, get stakeholders on board and then lose control • Get blamed if it doesn't work well • Carry the majority of the cost

to suit individual situations and skill is required to know where and when to apply these.

From consideration of the outcomes of the AWWA–AWA Honolulu Workshop and case studies, a number of key principles associated with involving stakeholders and achieving change have been identified. These are summarized in Table 3.2. The reader is encouraged to read the case studies at the end of the chapter or, even better, the original paper.

Hoppe et al. (2004) point out the need for development of social trust as an effective tool for preventing rancorous conflict. The range and depth of trust depends largely on the performance and behavior of institutions and organisations. Kasperson et al. (1992) identified four key social trust dimensions:

1. Commitment — to act objectively, fairly and openly with a commitment to assure the free flow of social information
2. Competence — in technical, social and cultural spheres
3. Caring/empathy — perceptions of a caring attitude towards dependent individuals and communities
4. Predictability — fulfillment of expectations

Table 3.2 Key principles in stakeholder management

Principle	Attributes
Transparency/ openness	• Stealth campaigns don't work – you always get caught out • Concern that stakeholder consultations can be token efforts – they need to get credibility with stakeholders • Tell the truth • Recognize that you can't please everyone • Need for transparency must be in proportion to the need for community involvement
Trust	• Need early engagement • Use simple messages • Provide supporting data • Be prepared for a different product than you expected at the outset, i.e., have flexibility and contingencies but need to keep end point in mind, there can be no compromise on drinking water quality • Develop social trust (Hoppe et al. 2004)
Two-way communication	• Listen and act • If required, have a third party peer review • Risk communication is important
Commitment	• Organisation and its representatives must demonstrate strong commitment to the goals
Competence	• Everyone involved must be competent, to ensure credibility
Caring and empathy	• Sincere concern and empathy with stakeholders are essential — people can smell out a fraud
Predictability	• Consistency in action and deed is important — do not change with the wind

Other learnings gleaned from Honolulu workshop participants were:

- Watershed management should occur at various scales
- High-level integrated policy and strategies at federal, state and local levels are important in supporting local efforts and preventing other agencies working at cross purposes
- Local policy and action plans should be developed in consultation with other agencies and the watershed community
- Management processes should integrate ecological, social, cultural and economic objectives.

Adaptive Management

An adaptive management framework can be useful for involving key watershed stakeholders in a structured change process. Bosch et al. (2004) provide a checklist for making adaptive management work. The checklist was developed at a workshop of natural resource management and adaptive management practitioners.

Adaptive management is

- an effective approach when managing within uncertain environments,
- *learning by doing* (Walters and Holling 1990),
- treating management interventions as experiments for testing and improving problem focused management,
- a mechanism for knowledge building, and
- co-learning (including surprise learning).

Other useful adaptive management resources can be found at the British Columbia Forest Service Web site www.gov.bc.ca/for.

CASE STUDIES

These case studies and summaries are intended to provide a guide and pointer to practices that might work in watershed management situations.

Community-Based Social Marketing

Doug McKenzie-Mohr in his article *Quick Reference: Community-Based Social Marketing* (CBSM) sets out the key steps involved in community based social marketing. CBSM is based on research that shows behaviour change is most effectively achieved through initiatives delivered at the community level which focus on removing barriers to an activity, while simultaneously enhancing the activity's benefits. It involves four steps:

- Identifying the barriers and benefits to the activity
- Developing a strategy that utilizes "tools" that have been shown to be effective in changing behaviour
- Piloting the strategy
- Evaluating it once it has been implemented across the community.

Stream Rehabilitation

Rutherfurd et al. (2000) set out a sound, 12-step, stream rehabilitation procedure, which can be adapted to watershed management planning. In summary, this involves:

What needs doing?

- Develop a vision
- Share the vision
- Describe the stream situation
- Identify assets and problems

Narrowing it down

- Set priorities
- Develop strategies
- Set measurable objectives
- Check feasibility

Doing it

- Design the details
- Plan the evaluation
- Schedule and supervise the works
- Assess and maintain the project

The principles translate well from stream rehabilitation to watershed management. Similar processes are set out in Koehn et al. (2001) and Bennett et al. (2002).

Australian Drinking Water Guidelines

The 2004 *Australian Drinking Water Guidelines* (ADWG) have been developed by the National Health and Medical Research Council (NHMRC) in collaboration with the Natural Resource Management Ministerial Council (NRMMC). The ADWG incorporates the *Framework for the Management of Drinking Water Quality* and provides the Australian community and the water supply industry with guidance on what constitutes good quality drinking water.

Several aspects of drinking water quality management require involvement with other agencies. For example, collaboration with the appropriate agency is necessary where watersheds and source waters are beyond the drinking water supplier's jurisdiction. Similarly, consultation with relevant health and other regulatory authorities is necessary for establishing many elements of drinking water quality management, such as monitoring and reporting requirements, emergency response plans and communication strategies.

The range of agencies involved in individual water supply systems will vary depending on local organisational and institutional arrangements. Agencies may include:

- Health and environment protection authorities
- Watershed and water resource management agencies
- Local government and planning authorities
- Nongovernment organisations
- Community-based groups
- Industry associations.

An integrated management approach with collaboration from all relevant agencies is essential for effective drinking water quality management. All major stakeholders that could affect (e.g., regulators, watershed boards) or be affected by (e.g., consumers, industry, plumbers) decisions or activities of the drinking water supplier should be identified. The list of stakeholders should be regularly updated.

The various agencies involved should be encouraged to define their accountabilities and responsibilities to support the drinking water supplier and, where appropriate, to coordinate their planning and management activities. Appropriate mechanisms and documentation should be established for stakeholder commitment and involvement. This may include establishing working groups, committees or task forces, with appropriate representatives, and development of partnership agreements, including signed memoranda of understanding.

USEPA Lessons Learned

The USEPA has documented a series of lessons learned from the pursuit of a watershed approach to protecting lakes, rivers, wetlands, estuaries and streams. The list is an attempt to identify the top lessons and present them in one place.

Top 10 Watershed Lessons

1. The best plans have clear visions, goals, and action items — These elements are folded into an implementation plan
2. Good leaders are committed and empower others — Leadership is a critical factor in making the watershed approach work
3. Having a coordinator at the watershed level is desirable — a watershed approach involves many stakeholders and requires a lot of coordination. A good coordinator is key to a successful watershed protection effort.
4. Environmental, economic, and social values are compatible — and are the foundation of a sustainable approach to watershed management
5. Plans only succeed if implemented — the greatest challenge associated with watershed management is to ensure that plans are actually implemented and do not gather dust on a shelf
6. Partnerships equal power — watershed management is about partnerships
7. Good tools are available — tools can include geographic information systems, watershed models, monitoring programs, best practice and "how-to" guides and funding assistance
8. Measure, communicate, and account for progress — appropriate performance measures keep issues on people's radar and allow stakeholders to share successes and to highlight new challenges
9. Education and involvement drive action — awareness and education programs can help people understand the problems and how they can be remedied
10. Build on small successes — start small and demonstrate success before working on a larger scale.

USEPA's *Top 10 Watershed Lessons Learned* booklet, available online at www.epa.gov/owow/lessons, includes a large number of useful case studies and links to key contacts and resources.

Principles for the Achievement of Sustainable Resource Use

Bellamy et al. (2002) were commissioned by the Murray-Darling Basin Commission (Australia) to provide an integrated overview of state reviews of integrated catchment (watershed) management (ICM) from across Australia and an analysis of potential learning of ICM in the Murray-Darling Basin. They present a set of principles for the future evolution of ICM.

Fostering institutional arrangements that are enabling

- Institutional arrangements that suit, and evolve with, their contexts
- Institutional arrangements that empower collaborative governance
- Institutional arrangements that empower integration with other governance systems
- Institutional arrangements that enable achievement of outcomes

Ensuring the integrity of participation

- Aim for inclusiveness
- Recognize and adapt to differences in stakeholder cultures
- Focus on empowerment, but beware of participant fatigue
- Use mixed modes of participation

- Look to build ownership, commitment and enthusiasm
- Reach agreement
- Foster fairness

Building individual, community and agency capacities

- Build capacities of all stakeholders — not just community — to participate
- Build collective capacities as well as individual capacities
- Build sectoral capacities to participate
- Broaden the experience base beyond current members

Moving beyond planning to implementation

- A cyclical strategic process
- Enable implementation
- Celebrate achievements

Turning towards adaptive management

- View systems beyond their parts
- Foster systemic learning

Focusing on achieving agreed outcomes

- Ongoing monitoring and review of progress
- Clarify accountabilities

Developing socially robust knowledge

- Understanding is evolving
- Coming to collective interpretations
- A broad focus on knowledge
- Knowledge sharing

Ecological Risk Management Framework for the Irrigation Industry

Hart et al. (2005) document an ecological risk management framework for the irrigation industry. The objective of the framework is to provide a robust process that will assist the irrigation industry to incorporate a transparent, scientific, precautionary and ecologically sustainable approach to its management of environmental risks. An important requirement of the framework is stakeholder involvement. Their report provides a useful overview of stakeholder involvement.

Stakeholders are those people or organisations who may affect, be affected by, or perceive themselves to be affected by a decision, activity or risk.

Broad stakeholder involvement is essential for a successful outcome in the types of ecological risk assessments considered in this framework, i.e., watershed-based and involving multiple hazards and multiple issues. Ideally, the stakeholder involvement should occur throughout the assessment process and involve both technical and non-technical people and organisations.

Although most ecological risk assessments now involve key stakeholders to some extent, Hart et al. believe this is often poorly done. Some of the problems they have observed include:

- Failure to involve important, but often 'difficult', individuals or groups of stakeholders. This leads to these groups questioning the credibility of the process (often through the press)
- Failure to include non-technical stakeholders (e.g., farmers and other landholders)
- Failure to take notice of legitimate issues brought forward by stakeholders
- Failure to account for long-standing disagreements between competing stakeholder groups
- Failure to put the necessary effort into informing and running the stakeholder workshops, resulting in key groups being disillusioned with the process and feeling alienated
- Failure to allow the stakeholders to drive the identification of the issues or hazards that need to be considered (e.g., where the issues have been largely decided prior to the involvement of stakeholders).

If the nature of a problem is such that risks are technically driven, there are ample data, and scientific understanding of the processes is good, risks may be assumed to be largely technical. In this case, technical risk analysis, without substantial sociological input, may be sufficient. Public participation may be treated with post hoc devices such as a public comment phase. However, when data are poor and there are substantial social pressures around the issues, greater public participation is warranted. In these circumstances, the solution is to involve those affected by the outcomes of risk assessments closely and continuously in the risk assessment process, making a marriage of the technical and social dimensions of risk. Non-technical information should enter at the planning stage, and non-technical participants should contribute to all elements of the risk assessment including the final outcomes of the study.

Stakeholder Map

It has been found that the preparation of a stakeholder map is a very useful way to ensure that all key stakeholders are included (or at least identified) and for understanding the socio-political settings (Freudenburg 1999; Glicken 1999, 2000). Stakeholder maps identify the key stakeholders (e.g., government organisations, scientists, residents, nongovernment organisations, industry representatives, traditional owners, unions and consumer groups) and map the linkages between them. There are a number of methods for acquiring stakeholder information including using existing data (archival and agency information), first-hand data collection (surveys, targeted interviews), and methods to identify gaps and oversights (sensitivity analyses, interdisciplinary double-checks and public involvement techniques). It is important to recognise that this process can take some considerable time.

Elicitation of Stakeholder Information

There are many methods available for eliciting information from stakeholder and community groups (Borsuk et al. 2001). They include preparation of stakeholder maps (Glicken 1999, 2000), surveys, targeted interviews through one-on-one contact, use of focus groups and stakeholder workshops, adaptive management (Walters 1986), building of Bayesian networks (Bacon et al. 2002a,b), multi-criteria decision-analysis (Roy 1999) and management strategy evaluation (Smith 1994).

The use of stakeholder workshops, particularly for involving key stakeholders in the early stages of the risk assessment and management process, is also covered.

WHAT SHOULD WE KNOW?

Review of the material presented above raises a couple of questions about future needs. These can be summarized as:

1. Making better use of the techniques and knowledge we already have. This chapter has shown that there is already a good body of knowledge about principles and frameworks and yet, in many cases they are ignored, or applied poorly; and we end up with sub optimal outcomes. A useful future need is to ensure that those in the water industry involved in stakeholder engagement and seeking change utilize best practice methods.
2. The watershed management industry needs to develop and maintain a solid community of practice for involving stakeholders and achieving change. This might be as simple as water associations fostering special interest groups on the subject or developing links with relevant professional associations (if they exist).
3. Keeping abreast of current stakeholder involvement and change management research, thinking and case studies.
4. Are the right people involved in the process? Stakeholder engagement can often be an add-on to a technical or engineering project and is run by the project managers as a distraction to the main focus of delivering the project. Project managers need to recognize that stakeholder involvement and change management are specialized fields requiring the involvement of specialists as part of the project delivery team.

OTHER RESOURCES

There is a vast amount of readily accessible and available resources on achieving change in natural resource (including watershed) management. Some useful links are below.

Citizen Science Toolbox

The Citizen Science Toolbox was developed in Australia by the Coastal CRC in partnership with Griffith University and the Department of Natural Resources, Mines and Water (Queensland). At the close of the Coastal CRC the toolbox was renamed and is now hosted at Griffith University as the URP Toolbox (https://www3.secure.griffith.edu.au/03/toolbox).

The Toolbox is a free resource of principles and strategies to enhance meaningful stakeholder involvement in decision-making. Stakeholders include not only communities, but also scientists and decision-makers. Meaningful involvement of all stakeholders occurs through a commitment to social learning — learning on the part of communities, other stakeholders and institutions.

One of the key elements of using the Toolbox is that strategies need to be tailored to fit unique issues and stakeholders. In many cases, a combination of different tools is required to effectively involve all stakeholders in decision-making. Included in the Toolbox are:

- Over 60 community involvement tools, from public meetings to consensus conferences
- Case studies of the uses of various tools and the reflections of stakeholders who participated
- An annotated bibliography of over 500 citizen science references
- Theoretical discussions of citizen science issues.

While the Toolbox has a focus on coastal and watershed environments, the principles and tools can be used in many other areas for a wide range of issues.

Web Sites

www.cbnrm.net/
CBNRM Net — The Community-Based Natural Resource Management Network

www.iap2.org/
The International Association for Public Participation, working through it members, helps organizations and communities around the world improve their decisions by involving those people who are affected by those decisions.

www.adaptivemanagement.net/
The Collaborative Adaptive Management Network develops best practices for solving complex conservation problems by creating and supporting adaptive and collaborative learning opportunities for practitioners.

REFERENCES

Bacon PJ, Cain JD and Howard DC. 2002. 'Belief network models of land manager decisions and land use change', *Journal of Environmental Management,* vol. 65, no. 1, pp. 1–23.

Bacon PJ, Cain JD, Kozakiewicz MB and Liro A. 2002. 'Promoting more sustainable rural land use and development: a case study in Eastern Europe using Bayesian network models', *J. Environ. Assessment Policy & Management,* vol. 4, no. 2, pp. 199–240.

Bellamy J, Ross H, Ewing S & Meppem T. 2002. *Integrated Catchment Management: Learning from the Australian Experience for the Murray Darling Basin,* Overview Report, CSIRO Sustainable Ecosystems, Canberra.

Bennett J, Sanders N, Moulton D, Phillips N, Lukacs G, Walker K, and Redfern F. 2002. *Guidelines for Protecting Australian Waterways,* Land and Water Australia.

Borsuk ME, Higdon D, Stow CA and Reckhow KH. 2001. 'A Bayesian hierarchical model to predict benthic oxygen demand from organic matter loading in estuaries and coastal zones,' *Ecological Modeling,* vol. 143, pp. 165–181.

Bosch O, Ross H, Witt B, and Smith C. 2004. *Making Adaptive Management Work: A checklist for*

Institutionalising Adaptive Management, Report for the Murray Darling Basin Commission, School of Natural and Rural Systems Management, University of Queensland.

Freudenburg WR. 1999. 'Tools for understanding the socioeconomic and political settings for environmental decision making,' in: *Tools to aid environmental decision making,* Dale VH and English MR (eds), Springer, New York, pp. 94–125.

Glicken J. 1999. 'Effective public involvement in public decisions', *Science Communication,* vol. 20, no. 3, pp. 298–327.

Glicken J. 2000. 'Getting stakeholder participation "right": a discussion of participatory processes and pitfalls', *Environmental Science & Policy,* vol. 3, pp. 395–310.

Hart B, Burgman M, Webb A, Allison G, Chapman M, Duivenvoorden L, Feehan P, Grace M, Lund M, Pollino C, Carey J, and McCrea A. 2005. *Ecological Risk Management Framework for the Irrigation Industry,* Report for the National Program for Sustainable Irrigation, Water Studies Centre, Monash University.

Hoppe, P with Rickson R, Burch D and Tufts-Rickson S. 2004. *Toward a collaborative model of industry/community relationships in environmental management,* Cooperative Research Centre for Coastal Zone, Estuary and Waterway Management, Australia.

Kasperson R, Golding D, and Tuler S. 1992. 'Social distrust as a factor in siting hazardous facilities and communicating risks', *J of Social Issues,* vol. 48, no. 4, pp. 161–187.

Koehn JD, Brierley GJ, Cant BL, and Lucas AM. 2001. *River Restoration Framework,* Occasional Paper 01/01, Land and Water Australia.

McKenzie-Mohr, D. *Quick reference: Community Based Social Marketing,* Online www.cbsm.com.

NHMRC, NRMMC. 2004. *Australian Drinking Water Guidelines,* National Water Quality Management Strategy, National Health and Medical Research Council.

Roy B. 1999. 'Decision-aiding today: What should we expect?' In: *Multicriteria Decision Making: Advances in MCDM Models, Algorithms, Theory and Applications.* Gal T, Stewart TJ and Hanne T (eds), Kluwer Publishers, Boston.

Rutherfurd I, Erie K and Marsh N. 2000. *Rehabilitation Manual for Australian Streams* vols 1&2, Cooperative Research Centre for Catchment Hydrology, Land and Water Resources Research and Development Corporation.

Smith ADM. 1994. 'Management strategy evaluation — the light on the hill', in: *Population Dynamics for Fisheries Management.* Hancock DA (ed.), Proc. Australian Society for Fish Biology Workshop, Australian Society for Fish Biology, Perth, pp. 249–253.

USEPA, Top 10 Watershed lessons, www.epa.gov/owow/lessons/.

Walters CJ. 1986. *Adaptive Management of Renewable Resources,* Macmillan Publishing Company, New York.

Walters CJ and Holling CS. 1990. 'Large scale management experiments and learning by doing', *Ecology,* vol. 71, pp. 2060–2068.

(Internet links quoted in this chapter were correct as of 22 May 2008)

4 RISK MANAGEMENT

Dan Deere, Joyce Perdek Walling, Ariamalar Selvakumar, Bruce Whitehill, Rachael Miller

CONTEXT

Water management organisations have recognised the importance of a multi-barrier risk management approach to protecting drinking water quality from contaminants. Often, the term multiple barrier is used to refer to drinking water treatment barriers. However, the concept can readily be applied to source water protection and storage management as well as to treatment processes and plumbing controls.

The multi-barrier approach involves the minimising of water quality hazards at each of the barriers to the fullest extent possible, rather than relying on the effectiveness of a single barrier system to protect public health. Water supply systems are almost unique in that the product is continuously delivered to the consumer and cannot be interrupted without placing great stress on a community. In the case of a system failure or contamination the product cannot be recalled, hence normal control systems that require acceptance testing are not applicable. There are limited on-line monitoring process available but most test results are received long after the product has left the plant and has probably been consumed. Consequently, most testing should be used as a tool for detecting trend changes and monitoring the effectivness of each barrier.

The strength of the multi-barrier approach is that a failure of one barrier may be compensated by effective operation of the remaining barriers, minimising the likelihood of contaminants passing through the entire treatment system. It is now widely recognised that watershed protection is a critical first step in the multi-barrier approach to protecting bulk raw water quality.

Preventive measures in these barriers should be applied as close to the source as possible, with a focus on prevention in watersheds rather than sole reliance on downstream treatment processes.

Failure to manage water quality can have catastrophic impacts. Even if no illness is detected, the community's trust in its water supplier can be lost and takes years to repair.

Risk management involves two key steps: the assessment of risk and the act of giving effect to the controls that manage the identified risks. This chapter deals with each of these two key steps in turn.

Watershed risk assessment involves assessing the risks presented to drinking water quality that arise in a particular watershed. The results of the risk assessment (both the evaluated risk and the level of certainty of the result) are key to deciding on appropriate risk management actions and their prioritisation.

OPTIONS

There are two levels (depths) of approach used to assess risks to water quality and prioritise control actions. These are:

1. A quantitative risk assessment, which uses exposure and reference dose data to assess the impact of water quality hazards on human health or aspects of ecology
2. A qualitative approach which involves expert groups assessing water quality issues, either as contaminants, pollution sources or hazard events, and prioritising these issues from this assessment. Methodologies used include assessing the risk (usually involving the likelihood and consequences) of an event.

A quantitative assessment approach is not generally used for typical watershed risk assessments because of the lack of data and the prohibitive amount of analysis required. Therefore, at present, qualitative risk assessments can be considered to represent best practice, the subject of this book. Quantitative techniques can be considered to be leading edge practice, but are beyond the scope of this book.

Frameworks

Within the range of qualitative techniques that can be used it should be recognised that there is a range of established frameworks within which the qualitative assessment can be carried out. It is difficult to discriminate any one of these as being best practice, rather all appear to be best practices with the choice of framework that is truly 'best' depending on the specific circumstances.

Ecological Risk Assessment (ERA)

This method can be used to assess the level of risk to the health of river ecosystems from management actions. The ERA process has evolved as a tool in assessment of multiple stressors on complex ecosystems. Generally, the process covers the following steps:

- Problem formulation — establishes the goals, breadth and focus of the risk assessment involving consultation, examination of data, consideration of policy and regulatory issues and site specific factors
- Risk analysis and assessment — expert group(s) develop profiles of environmental exposure (likelihood) and ecological effects (consequences) of actions, events or land use changes
- Risk characterisation — exposure and effect profiles are integrated to provide an estimate of the level of risk (risk = likelihood × consequence). Results can be expressed qualitatively or quantitatively
- Risk management — ERA information is used to define the appropriate actions and highlight the priority factors to be addressed.

ERA is a useful broad framework to apply to multiple value, broad-scale watershed risk assessments. Often, the watershed risk assessment considers many ecological and social values, not just drinking water supply. Therefore, the ERA framework is recommended for undertaking broad risk assessments of multiple values in multiple use watersheds.

Risk Management

There are standards for risk management which set out a framework for qualitative risk assessment (for example see AS/NZS 4360:2004 Risk Management). The standard approach to risk management is not specific to watersheds, but is a best practice approach to risk management in general that can readily be applied to watersheds.

The steps involved are:

- Establish the context
- Identify risks
- Analyse risks
- Evaluate risks
- Treat risks
- Monitor and review risks.

This process is shown in Fig. 4.1.

Standardised approaches to risk management are useful if watershed risk

Figure 4.1 AS 4360 risk management process

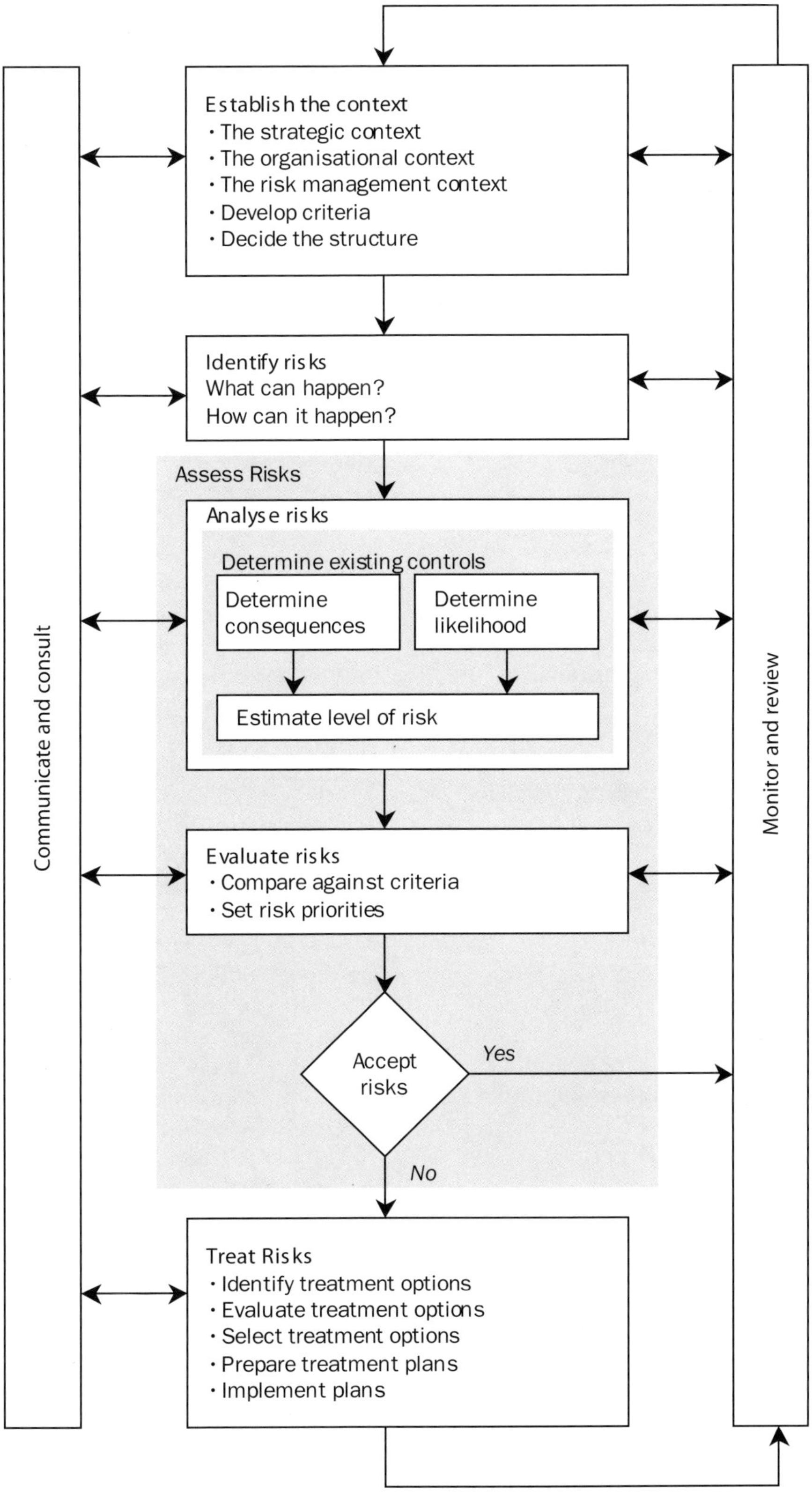

assessment is being undertaken as part of a broader organisational risk assessment. The advantage of the generic risk assessment and management framework is that the approach can be applied to the management of all business risks. The organisation can then compare risks to public health arising from watershed contamination of water supply to risks of occupational safety of workers, environmental impacts from sewerage asset failures and financial risks. Therefore, standard approaches to risk management are recommended for undertaking risk assessments of watersheds using the same approach, and for broad comparison, with many other organisational risks.

Hazard Analysis and Critical Control Points

Under the hazard analysis and critical control points (HACCP) process a team must assess potential risks to water quality. The focus of HACCP is to identify and implement actions at *critical control points* (CCP) which are activities, procedures or processes to which strict control can be applied and which are essential to prevent a hazard or reduce it to an acceptable level. The HACCP approach has the following principles:

1. List all potential hazards
   * Conduct a hazard analysis
   * Determine control measures
2. Determine CCPs
3. Establish critical limits for each CCP
4. Establish a monitoring system for each CCP
5. Establish corrective actions for deviations that may occur
6. Establish verification procedures
7. Establish record-keeping and documentation.

Control points are commonly found in manufacturing and treatment plants but rarely in natural open systems such as watersheds and water storages. Therefore, watershed management actions and programs are better described as 'Prerequisite Programs' within the contemporary application of HACCP, or as 'Operational Prerequisite Programs' within ISO 22000, which is essentially an ISO version of HACCP, expanded to create a management system.

A narrow interpretation of HACCP is not a particularly useful framework for watershed risk assessment since HACCP is primarily a process control tool applied to production processes, such as water treatment. However, the broader definition of HACCP, including the 'Prerequisite Programs', or ISO 220000 with its 'Operational Prerequisite Programs' are very amenable to application to watershed management. Both HACCP and ISO 22000 are generic food/beverage safety management systems which naturally apply to the production of potable water. Therefore, the broadly defined HACCP or ISO 22000 frameworks are recommended for undertaking risk assessments of drinking water watersheds where potable water supply is the focus of the assessment, rather than broader ecological outcomes.

World Health Organisation Water Safety Plans

The World Health Organisation (WHO) Guideline for Drinking Water Quality introduces a management framework for safe water. These Water Safety Plans (WSP) are a systematic approach to understanding the specific hazards relevant to a given water supply and to effective management of the processes most suitable for their control in the given system. WSPs have the following key elements:

- Health based targets — developed by WHO using quantitative microbial risk assessment
- System assessment — determines the system capability to meet health based targets
- Effective management — control measures — understanding the capabilities and limits of barriers
- Management plans — document system assessment and monitoring, describe actions to be taken during normal operation and incident conditions, including upgrade and improvement

and documentation and communication
- Public health surveillance — verifies that the elements of the WSP are operating properly, including auditing.

The WSP approach is very similar to HACCP except that the emotive terms 'critical' and 'prerequisite' have been omitted from CCPs leaving in their place 'Control Points' and 'Supporting Programs' (equivalent to the 'Critical Contol Points' and 'Prerequisite Programs' in HACCP). Watershed management programs are readily defined as Control Points or Supporting Programs, making the WSP approach intellectually much simpler to apply to watershed management than conventional HACCP. Note that HACCP is also largely a WHO program and WHO deliberately used a slightly different system when developing the WSP concept to avoid the difficulty of applying conventional CCPs to watersheds.

WSP is a useful drinking water supply risk management framework to apply to drinking water watersheds and logically would be preferred over HACCP and ISO 22000 because it is tailored for water supply. However, the broader familiarity of HACCP and ISO 22000, and the availability of certification for these more widely used systems means that many utilities have used HACCP and ISO 22000 systems, albeit being guided by the WSP approach.

One particular benefit of HACCP in watersheds that have significant agricultural stakeholders is that all food chain components are adopting, or have adopted, HACCP and/or ISO 22000. Therefore, communication about what is to be achieved for the water supply watershed is facilitated if such familiar terminology is used. The WSP framework is recommended for application to all watersheds, but the availability of certification for HACCP and ISO 22000, and the broader application, sometimes means that organisations adopt the latter. In practice, WSP and HACCP or ISO 22000 are very similar, the differences being largely semantic and in relation to certain points of detail.

BEST PRACTICE

Within the United States, New Zealand and Australia, the water industries have developed national frameworks that include water quality risk assessment and management. The New Zealand system is the 'Public Health Risk Management Plan'; the Australian system is the 'Framework for the Management of Drinking Water Quality'; and the US system is the 'QualServe' program administered by the American Water Works Association. All are essentially risk and quality management frameworks. The Australian and New Zealand examples are specific to drinking water quality, whereas the US one is business-wide, with water quality being just one objective.

The New Zealand and Australian systems are very similar to the WHO WSP, with the New Zealand system dropping the word 'Critical' as did WHO, although the Australian system is more HACCP-like, retaining the word 'Critical' and providing a very stringent definition of CCPs that rules out most watershed processes.

Within each of these three countries, the use of the relevant national framework can be considered a best practice approach and is the logical default framework within which to undertake a risk assessment and management plan. The suggested best practice risk assessment approach is any one of the above, applied where it best suits the entity undertaking the risk assessment. All have essentially the same risk assessment and management framework elements, albeit in different orders and with different jargon.

For the purposes of this chapter, we have identified the elements that make up a best practice qualitative risk assessment and risk management framework (illustrated in Figure 4.2).

Risk assessment involves three steps:
- Screening level risk assessment of water quality
- Hazard event risk assessment for supply systems
- Overall risk assessment combining water quality and hazard event risk results.

Figure 4.2 Best practice risk assessment and management framework elements

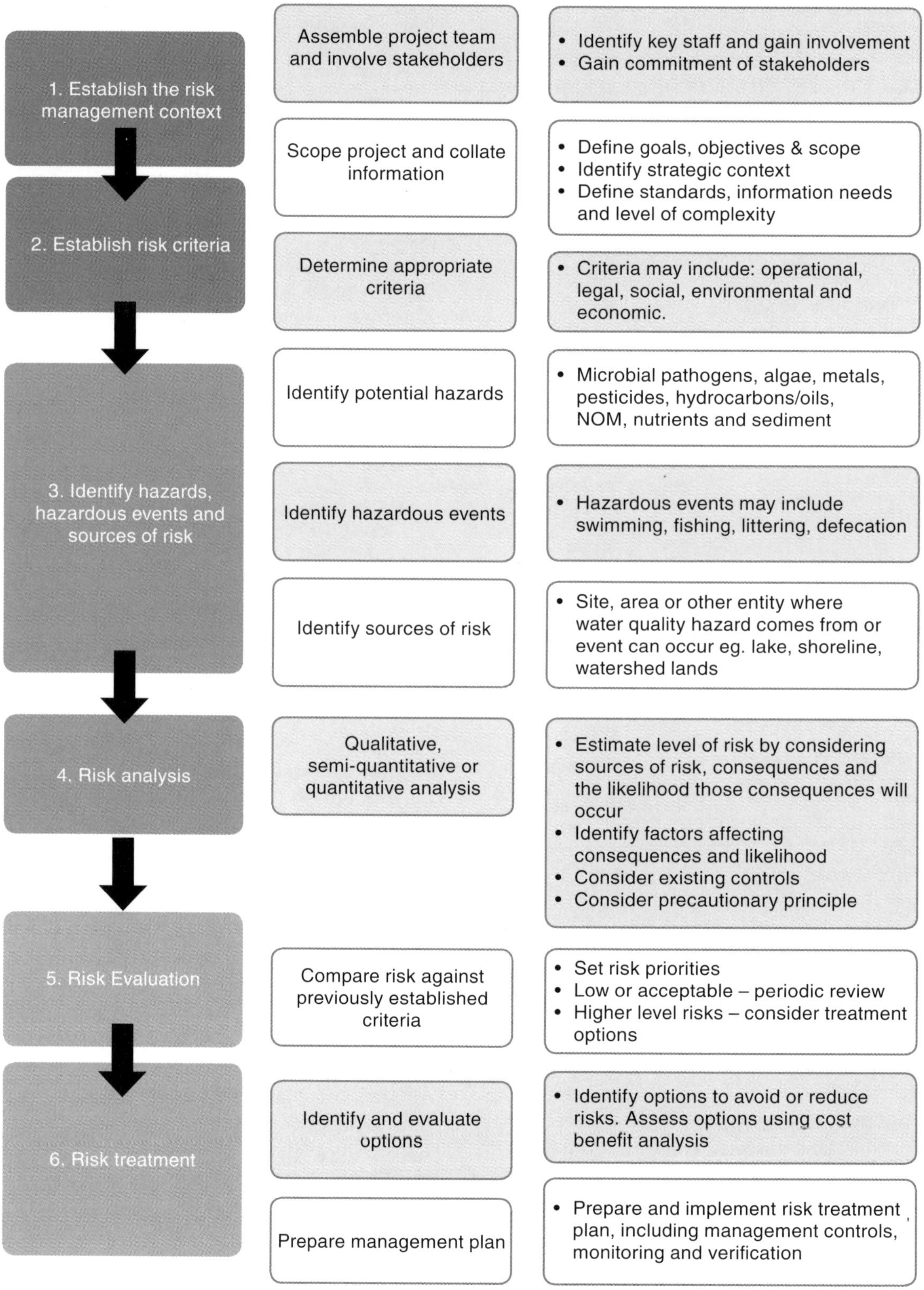

Step 1. Screening Level Risk Assessment of Water Quality—The screening level risk assessment involves a comparison of water quality with guideline or target values (quality or contaminant load) at selected endpoints. Modifiers can also be applied to the risk result to differentiate between human health and aesthetic impacts. This quick evaluation fails to identify risks from possible changes or incremental development but does identify risks currently being experienced to water supply where treatment systems might fail or be overwhelmed.

Step 2. Hazard Event Risk Assessment for Supply Systems—The hazard event risk assessment identifies and gives a general assessment for watershed events that cause or influence water quality hazards. This is usually undertaken using a workshop approach using an *expert group* to assess hazard events in the water supply systems.

Results from process mapping of water quality hazards can also be used as an input to the risk assessment methodology to guide the identification of hazard events and incorporate the significance of contributing factors such as pollutant loads into the hazard event risk assessment.

Step 3. Overall Risk Assessment—The screening level risk assessment results for water quality and hazard event risk results are combined to yield overall risk results for hazard events and to determine risk management (treatment) priorities.

Definitions

The various risk assessment approaches each define water quality hazards in different ways. Clearly presenting this definition is vital to setting the context of the risk assessment, and to identifying what event or entity requires risk management (treatment). In other words, risk of what and risk to whom? One method is to define water quality hazards as contaminants or water conditions that can cause harm to human health. These are represented by specific analytes listed in appropriate guidelines or supply agreements.

A circumstance or incident that can cause a water quality hazard either directly or indirectly is termed a *hazard event*. For example, a bypass from a sewage treatment plant or runoff from livestock industries is a hazard event. Clear description is required because this is the circumstance that will most likely require *risk management* (treatment).

A site, area, activity or other entity where a water quality hazard(s) comes from, or where a hazard event can occur is termed a *hazard source*.

Conceptual models or *process maps* can be used to better understand water quality hazards in the risk assessment process and serve as qualitative hypotheses that assist decision making. These describe the multiple physical, biological, and chemical factors in a system and their sources, and the pathways by which they are likely to impact multiple resources. Process maps can incorporate the relative importance of causes and control to the occurrence of the water quality hazards, and the certainty of this information.

Uncertainty in Risk Assessment

Risk assessments can be uncertain as a consequence of gaps in knowledge and understanding. Limitations to certainty regarding water quality risk assessment include:

- Limited understanding of fate and transport mechanisms which determine concentrations and duration of contaminants in the environment
- Characterising the responses of natural systems to contaminants and other stressors
- Lack of knowledge on in-stream processes and how they react to disturbance
- Limited data about the cumulative effects of contaminants
- Lack of knowledge of future land uses
- Lack of knowledge of the effect of incremental changes.

It is widely acknowledged that risk assessments should include some estimate of the certainty in the predictions, and a summary of the assumptions used. Characterising certainty can provide a better understanding of the limitations of risk assessments, and how these limitations can be reduced. Sensitivity analyses that examine variability of risk results due to changes in the evaluation criteria can also be used an indicator of certainty.

A Precautionary Approach

Certainty for risk assessment results can be considered in selecting the appropriate risk control to address identified hazards, ranging from preventative actions where the assessment is certain, to more precautionary actions where the assessment is less certain.

This precautionary approach is supported by the precautionary principle of ecologically sustainable development which states that 'if there are threats of serious or irreversible environmental damage, lack of full scientific certainty should not be used as a reason for postponing measures to prevent environmental degradation.' Serious or irreversible environmental damage includes impacts on human health through the bulk raw water supply.

Certainty results can also be used to prioritise an organisation's water quality monitoring and research programs.

Impact of Incremental Change

Most risk management techniques provide little guidance in relation to incremental change; those involving some form of monitoring will only measure the impact after it reaches a critical value, usually at a time when it is too late to reverse. There remains a need for improved approaches to capping the magnitude or density of impacts in a precautionary way.

Best practice should involve a structured and systematic approach to the management of drinking water quality from watershed to consumer, to ensure its safety and reliability. This may include elements of HACCP, environmental and risk management processes but must apply them in a drinking water supply context to support consistent and comprehensive implementation by the water industry.

Controls

The implementation of controls is required upon the completion of the assessment of the risks to the drinking water supply system (generally using a generic risk management process). The steps involved in implementing controls typically include the following:

- Evaluation of preventive measures for drinking water quality management
- Establishment of operational procedures and process controls
- Verification of drinking water quality
- Management of incidents and emergencies.

Furthermore, the risk management actions need to be underpinned by management commitment, stakeholder involvement, emergency response, employee training, community consultation and research and development. Watershed management objectives cannot be achieved without the participation and support of the watershed community, regulators, local government and landholders.

The challenge for a watershed manager is to organise a variety of watershed management instruments and information sources into a structured framework or plan to address hazards to water quality and watershed health. Watershed management initiatives include:

- Regulatory enforcement using pollution control powers and guidelines
- Land-use control using planning powers of other agencies (water protection zoning in planning controls and control/prohibition of inappropriate development)
- Incentives and assistance programs (e.g., riparian area protection)
- Education programs (e.g., school programs, ranger based tours and informative brochures)
- Influencing federal, state and local government policies to support watershed management.

These initiatives are supported by:

- Watershed information
- Modelling and water quality data
- Agreements, liaison and goodwill with other government and community stakeholders.

The management initiatives can be presented as a hierarchy of instruments (Fig. 4.3), from persuasive instruments at the base to punitive instruments such as regulatory enforcement at the peak of the pyramid.

The choice of instruments is not static, and must be able to respond to changes in watershed condition or as knowledge of watershed processes increases. Where appropriate, combinations of instruments that complement each other should also be preferred to the use of instruments in isolation. The issue of stakeholder engagement is covered in Chapter 3.

The factors to consider in choosing watershed protection instruments include:

- Scientific certainty about the impact on water quality and watershed health
- The risk of impact on water quality and watershed health
- Imminence of the impact on water quality and watershed health
- Level of cooperation of industry/regulated party
- Clarity of influence or jurisdiction on a watershed pressure.

Figure 4.3 Pyramid structure of watershed protection instruments

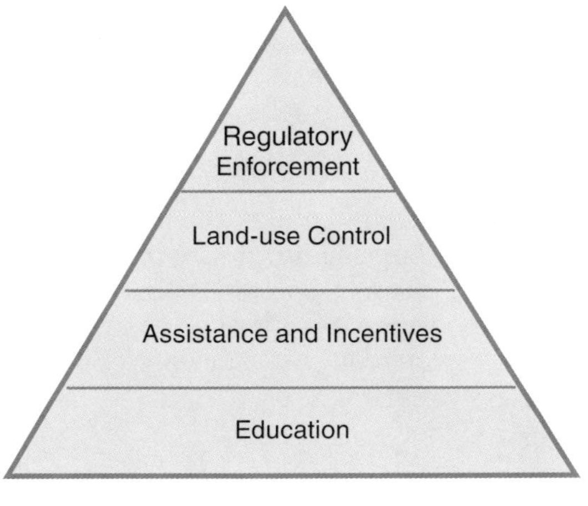

Protecting Watershed Health

Watershed health in its entirety is a complex issue. It involves factors including the provision of environmental flows, protection of river water quality, aquatic biodiversity and habitat and management of native vegetation and land.

Watershed health differs from the integrity of selected ecological indicators; it implies a viable condition, a self-sustaining state or series of states, which are compatible with a human use. Compatible human use can vary according to the objective of the investigation, assessment or plan that is addressing the issue.

For the purposes of watershed management for water supply, watershed health can be defined as the condition of aspects of the environment that promote water quality. Actions to protect these environmental aspects can therefore regarded as key *risk controls*.

Potential long term changes in land use must also be considered.

Riparian Zones and In-channel Processes

Riparian zones and in-channel processes are examples of environmental aspects that influence water quality hazards.

Riparian zones include lands directly adjacent to water bodies such as lakes, reservoirs, rivers, streams and wetlands. The zone represents the interface between watershed lands and the water conduits in water supply systems and are important control points for water quality management. River bank vegetation can act as a buffer along rivers and reservoirs. Recent (2004) research by the Cooperative Research Center for Water Quality and Treatment in Australia has confirmed previous studies proving the value of these areas in retaining or transform diffuse source pollutants. In addition, riparian areas also produce a more favourable environment for aquatic ecosystem and in-channel processes.

In-channel processes include the geomorphic aspects of watershed creeks and rivers and the aquatic flora and fauna that live there. These processes may remove or inactivate water quality hazards. For example, natural predators that live in water can consume microbial pathogens, and natural organic matter can be degraded and metabolised by river biota.

Watershed management and land-care groups often have riparian protection programs in watersheds for other protection purposes such as salinity management and wildlife corridors. There is an opportunity for water utilities to link to these existing programs and integrate drinking water quality protection.

Working with Industry and Stakeholders

Integration of water quality protection into industry codes of practice is another effective mechanism. For example, codes of practice for timber harvesting in water watersheds and management of domestic, on-site wastewater treatment systems have been amended to incorporate special requirements for activities in watershed areas. This option requires goodwill from the industry and a willingness to enforce these requirements, which is often lacking (see Chapter 3).

Another option for transient activities is to look for alternative areas. This may mean relocation or development of facilities for discretionary activities such as recreational swimming, fishing, and four-wheel driving (see Chapter 8).

Water Protection Planning Zones

Recognising drinking water quality protection in land-use planning and zoning is a very effective mechanism. Compatible land uses are then regulated through planning processes implemented by regional and local town planning schemes.

This is particularly effective if it allows for varying levels of protection. A good example of this is having a tiered zoning tool for strategic or vulnerable areas of a watershed or groundwater area, grading down to less critical protection areas. For example, the Department of the Environment in Western Australia has effectively worked with the planning regulator to achieve recognition of a priority 1 and 2 system in the state planning scheme. Priority 1 is termed 'Water Catchments Reservation', and is based on risk avoidance with the key goal of eventually being under government/water supplier ownership. Priority 2 land is termed 'Rural Water Protection Zone', based on risk minimisation and is compatible with low intensity rural land use activities.

In Victoria (Australia) four adjacent water authorities have adopted a common policy for assessing development applications that is transparent and publically available.

Cost-benefit Analysis of Risk Treatment Options

Cost-benefit analysis (CBA) is a test for the economic merit of a proposed investment (project) to assess its ability to deliver a (medium to long term) future benefit stream sufficient to outweigh the project's current and future costs. The usual methodology is to compare the with project and without project cases. Information for CBA comes from risk assessment, and cost estimates. CBA is a vital step in prioritising proposed risk management (treatment) actions to determine the likely net benefits of a program of risk control actions.

Unfortunately CBA is rarely done well in the risk assessment/management process. The step of choosing risk control actions is not usually conducted as a discrete step of the risk management process, normally being combined with the risk assessment and evaluation step. If watershed management is to maintain legitimacy as an option for water quality protection in the multi-barrier approach, then a transparent CBA is required for the associated actions.

CASE STUDIES

Water Corporation (Western Australia)

This section provides a description of the key steps followed in undertaking a risk management process in drinking water watersheds and storages in an Australian context, using Water Corporation of Western Australia as the example. The approach can be used to assist decisions on access in relation to particular activities for particular locations in the watershed or storages.

Due to the variability in watersheds and storages, this provides a general guide only. It is not intended as a detailed 'how to', but rather a summary of key issues that should be considered when undertaking a risk assessment. The key steps in the Australian Risk Management Standard (AS/NZS 4360:2004) comprise:

Risk Management Context — Establishes the goals, objectives, strategies scope and parameters of the activity that is the subject of the risk management process.

Risk Identification — Identifying hazardous events and sources of risks —what can happen and how?

Risk Analysis — Consideration of the sources of risks, their consequences and the likelihood those consequences will occur, within the context of existing controls.

Risk Evaluation — Comparing the level of risk found during the analysis against previously established criteria. The output of a risk evaluation is a prioritised list of risks for further action.

Risk Treatment — Identification of the range of options for treating risks, assessment of those options and preparation and implementation of risk treatment plans.

Monitor and Review, Communicate and Consult — The first step in undertaking a risk management process is to establish the strategic, organisational and risk management context within which the assessment is to take place. Key elements are:

Establish Project Management Structure and Responsibilities — Depending on the level of complexity of the risk assessment, this may involve the appointment of a single project manager and a project team with appropriate knowledge and expertise and/or the appointment of a specialist risk analyst. It is important that personnel with subject matter expertise and knowledge of the system or activity under review are involved in the risk management process.

Stakeholder Involvement — The Australian Drinking Water Guidelines (ADWG) Framework for Drinking Water Quality Management (NHMRC & NRMMC 2004) notes that stakeholder involvement is vital in drinking water quality management. This is particularly important when making decisions where there can be multiple interest groups, often with conflicting attitudes and values. Decisions about activities in drinking water watersheds and storages raise broader questions about the relative value placed by the community on the social, environmental and economic benefits and costs associated with activities in these areas.

A key aspect of the risk assessment process is communication with internal and external stakeholders (AS/NZS 4360; 2004). It is important to involve key stakeholders throughout the risk assessment process so that those responsible for making decisions about watershed activities, and those affected by the decisions, understand the basis on which decisions are made, and why particular management actions or controls are required.

Risk Assessment

Risk assessments are used by water utilities in water quality management for a number of purposes. Those include day-to-day risk minimisation such as for operational planning, plus also higher level risk prioritisation for planning, and then to define the appropriate solutions for longer term risk mitigation. The application typically defines the level of detail integrated into the risk assessment and the type of risk assessment, whether it be qualitative or quantitative.

Limitations to the selection of the type of risk assessment are usually based on the type of quantitative data that can be put into the risk assessment. Generally, it is now becoming evident that, for higher level source planning purposes, qualitative risk assessments are not the most effective. These risk assessments deliver a qualitative result, but have a high level of quantitative information *built in* to the risk assessment. For operational purposes, semi-quantitative risk assessments allow for the easy prioritisation of operational resources (surveillance, sampling, etc.)

The mechanism by which the risk assessment is applied is through a corporate standard or manual, titled the *Source Protection Operations Manual* (SPOM). This requires responsible staff to routinely review risk assessments for sources operated by the organisation, and then determine appropriate levels of on-ground operational actions that are required to mitigate that risk. These actions specifically include surveillance (rangers), site inspections, water quality monitoring and installing signage. Overall, though, all these actions also contribute to knowledge gathering of the characteristics of a source and, as such, provide a key understanding of how best to operate that water supply system most effectively, from watershed to tap/faucet.

The risk framework adopted for source protection by the Water Corporation was based on the key requirements of the ADWG Framework for Drinking Water Quality Management, as well as Australian Standard 4360. This was owed to a corporate commitment to implementing the ADWG Framework as well as to address duty of care issues with drinking water supply and operation. The Water Corporation has been delegated some on-ground operational source protection roles from the regulator.

An example of the risk assessment framework is outlined in Figure 4.1. Key features of the approach include recognition of factors affecting likelihood of the risk occurring, residual risk (remaining risk after barriers have been recognised) and a sanitary survey score for microbiological risks.

The standard specifies that higher levels of risk identified by the risk assessment require more thorough on-ground operations. An example of this is a source water risk assessment that has identified transient access to the waterbody by cattle from a neighbouring paddock. The required operational response to this risk, dictated by SPOM, is at two levels:

- Identify the immediate risk posed to the raw water quality which can then be used to adjust downstream operations if required (such as intensive monitoring, source shut-off, emergency treatment or reporting to regulator)
- Investigate access conditions and, where possible, rectify issue by ceasing access.

In regard to the first objective, SPOM requires the following:

- Increase site visits to the location to better understand the locations and times of access and cattle numbers
- Increasing raw water quality monitoring routinely as well as in storm events or during known access times.

Once information is collated, raw water quality results must be reported to a central branch as well as local downstream operators where decisions can be made on the appropriate downstream operational response. In regard to the second objective, SPOM requires the following:

- Increase site visits to the location to better understand the locations and times of access and cattle numbers
- Identify damaged fences (and ownership) that are allowing access
- Contact landowner and request the repair of fences; or provide assistance
- Follow up later to ensure actions have been completed and access has ended.

United States Regulatory Framework

Drinking water source protection extends beyond controlling individual sources of contamination to address problems and solutions on a regional or watershed basis. In the United States, the Safe Drinking Water Act Amendments of 1996 require state drinking water agencies to examine the sources of drinking water as an additional layer of protection for drinking water supplies. States are required to establish source water assessment programs which must include the following four major elements:

- Delineating (or mapping) the source water assessment area
- Conducting an inventory of potential sources of contamination in the delineated area
- Determining the susceptibility of the water supply to those contamination areas
- Releasing the results of the determinations to the public (USEPA 2000).

These assessments are designed to provide a thorough understanding of the potential risks to each water supply and serve as the basis for making risk-based watershed management decisions.

Risk-Based Watershed Management

After the risks in the water supply area watersheds are identified and characterized in terms of public health impacts, the risk management process begins. Risk-based watershed management consists of:

- Making decisions on management activities that guide both short-term and long-term water quality objectives
- Implementing management activities in an adaptive manner to continually improve understanding, practice, and results (Crowther and Partners 1999).

These management activities are 'the planned manipulation of one or more factors of the natural or disturbed drainage as to effect a desired change in or maintain a desired condition of the water resource' (Black 1996). Challenges to watershed management include diffuse pollutant sources and geographical boundaries that encompass multiple municipalities, provinces, or states.

A risk-based decision approach maximizes the chances of success by recognizing the existence of multiple objectives from the start, involving and informing stakeholders, structuring the decision process, and striving to reach a consensus solution which provides the best combination of outcomes on all objectives (Butcher 1999). Achieving objectives is likely to include a multi-barrier approach, consisting of minimizing threats to water quality at each step in the water delivery process. This multiple-barrier approach goes beyond the traditional emphasis on treatment to address new challenges and reflects a better understanding of the need for a coordinated source water protection or watershed management effort. The multiple-barrier approach encompasses delineation and prevention of contamination of drinking water sources, treatment appropriate to the quality of the source water, well-engineered distribution and storage systems, operator training and certification and an informed and involved public. Since 1996, a multiple barrier approach has been promoted in the United States (USEPA 2005a) and watershed management is a critical step in this approach.

Watershed management begins with identifying and prioritizing stakeholder objectives, such as meeting specified water quality parameters. Stakeholders include regulatory agencies, local government, landholders, and the watershed community. By involving stakeholders, there is a greater chance that the problem will be successfully defined as they contribute knowledge about the situation in and around the watershed and serve an important role in making the issue as part of their agenda. The next step is identifying potential management options which include best management practices (BMPs) for stormwater runoff and upgrades to industrial and municipal wastewater treatment

Table 4.1 Structural and non-structural BMPs

Structural BMPs	Non-structural BMPs
• Dry Ponds (Detention) • Wet Ponds (Retention) • Grass Swales (Wet/Dry) • Filter Strips • Constructed Wetlands • Bioretention Cells • Infiltration Trenches • Infiltration Basins/Porous Pavements • Green Roofs	• Public Education and Outreach • Better Site Design • Dumping Controls • Vegetation Controls • Reduction/Disconnection of Impervious Areas • Alternative Product Substitution • Street and Storm Drain Maintenance • Spill Prevention and Cleanup • Illegal/Illicit Connection Control • Stormwater Reuse

plants. BMPs encompass a wide range of approaches, both structural and non-structural, as shown in Table 4.1. Structural BMPs are engineered systems and methods designed to provide temporary storage and treatment of runoff for the removal of pollutants. Non-structural or source control BMPs are practices that prevent pollution by reducing potential pollutants at their source before they come into contact with water (Field et al. 2005). Management options identified need to be evaluated for their ability to reduce contamination. In many cases, watershed and water quality models are used to evaluate and compare management alternatives. Models are also useful tools for examining options used concurrently and for linking pollutant discharge amounts to concentrations in the water body (see Chapter 10). Once measures are implemented, water quality monitoring and other data collected will assess their effects on the water body.

An adaptive approach to implementation provides for the opportunity to reevaluate and adjust management measures based on results. Adaptive management is a systematic approach to improving understanding of public health impacts and ecosystem functioning and management performance by learning from the outcomes of management policies and practices. Adaptive management involves an iterative six-step process:

1. *Assess* the problem by identifying key uncertainties and generating preliminary hypotheses
2. *Design* management actions as a means of resolving or reducing uncertainty
3. *Implement* management actions
4. *Monitor* results focusing on a set of key ecosystem indicators
5. *Evaluate* outcomes against original designs and hypotheses
6. *Adjust* implementation strategies, in response to monitoring and evaluation results (Crowther and Partners 1999).

Growing numbers of drinking water utilities throughout the world are engaged in successful watershed management programs. Their solutions include but are not limited to:

- Establishing regulations to prohibit or restrict land uses that may release contaminants in critical source water areas
- Educating and encouraging citizens and businesses to recycle used oil, limit their use of pesticides, participate in watershed cleanup activities, and other prevention activities
- Purchasing land or creating conservation easements for protection zones near drinking water sources
- Developing partnerships between the water utility and local governments
- Engaging with local farmers to implement best management practices aimed at protecting sources of drinking water (USEPA 1999).

The following case studies are examples of how such watershed management scenarios are determined and implemented. While none can directly apply to another watershed situation, some pieces may help to develop watershed management plans.

US Case Study 1: Managing Urban Stormwater, Massachusetts Water Resources Authority, Boston, Massachusetts (USEPA 1999)

- Primary Sources of Water: Quabbin and Wachusett Reservoirs
- Watershed Area: Quabbin: 187 square miles; Wachusett: 110 square miles
- Population Served: 1,932,000

Drinking water source quality is often adversely affected by stormwater runoff. After a large rainfall, contaminant concentrations increase significantly, further stressing drinking water treatment facilities. Although the Massachusetts Water Resources Authority does not regulate stormwater releases from construction sites, the Metropolitan District Commission (MDC) Division of Watershed Management works with petitioners to review all plans for the design and construction of stormwater and erosion control projects. These control projects are required under the state's Watershed Protection Act as well as the Wetlands Protection Act.

Review Authority

The Massachusetts Watershed Protection Act requires division review of all proposed alterations in watershed areas that are within 400 feet of designated tributaries, wetlands, and flood plains. Proposals submitted to local conservation commissions are reviewed by division staff for water quality concerns during and after construction. Following their review, staff advise the commissions of their findings. During 1997, MDC division personnel reviewed 98 applications with an eye toward erosion and sediment control and water-quality issues. Each project's design ultimately conformed to watershed protection regulations and met stipulated water quality standards. Additionally, annual watershed sanitary surveys help MDC staff identify areas of concern. Once a specific threat to human health is identified, the MDC works with the responsible party to mitigate the situation. Several projects with the state highway department began in this way.

Watershed Approach to Stormwater Management

Besides these ongoing activities, the MDC has hired an environmental engineering firm to prepare an in-depth stormwater management plan for the entire watershed. Once completed, the plan will include a series of pollutant-loading analyses at the sub-basin level and recommended best management practices (BMPs). The Massachusetts Water Resources Authority and MDC plan to conduct workshops to help municipalities implement the BMPs, and may provide technical and financial assistance.

Community Ordinance Revision

The MDC Watershed Division also provides technical assistance to communities to help establish or revise erosion and sediment control bylaws and regulations, site plan bylaws, and subdivision regulations. Division staff are working with one community to develop model subdivision regulations for use with other communities. These regulations are intended to foster a more progressive approach that includes concerns for water quality.

US Case Study 2: Community Involvement, Burlington, Vermont (USEPA 2005b)

- Primary Sources of Water: Lake Champlain
- Population Served: 40,000

Burlington, Vermont, is a city of approximately 40,000 residents located on the shores of Lake Champlain. Not only is Burlington the largest city in Vermont, but it is also located in the most populous and fastest growing county in the state. In addition to being a popular source of recreation and commerce, the 120-mile long, 12-mile wide lake is the source of drinking water for Burlington and other waterfront municipalities; there are no public groundwater sources in the city. The watershed for the lake spans most of upstate Vermont, stretching from western Vermont into eastern New York State. As with other large resource areas that stretch across multiple jurisdictions, protection of the Lake Champlain watershed requires thoughtful coordination between federal, state, and local governments, agencies and individuals. Because the city discharges its treated wastewater and stormwater into Lake Champlain and takes its drinking water from the same source, the citizenry and businesses of Burlington have taken a strong interest in the lake's water quality. This has led to an exemplary drinking source water management and protection program.

Local Involvement and Developing the Protection Plan

Community involvement has been part and parcel of Burlington's source water protection program. Public forums held during the 1980s to address beach closure issues on Lake Champlain initiated much of the subsequent planning work. The upgrading of three Burlington wastewater treatment plants (WWTPs) was tied into a bond issue on a public ballot. Hearings to address the WWTP upgrades, extension of one WWTP outfall further into the lake, and the separation of combined sewer outfalls (CSOs) were well attended by the public and gained considerable news coverage. There was significant general public involvement in the upgrade to the drinking water treatment plant and in efforts to protect the area around the intake. For example, a broad representation of the public was involved in the extension of a wastewater outfall beyond the breakwater and Burlington Harbor. Representatives of interests from the perspectives of marinas to the preservation of historic sunken ship sites were involved in the successful location of the new outfall. Citizens have been involved through committees, surveys, and meetings at each juncture of the program's development. There will also be a public outreach component as part of the Englesby Brook Watershed Restoration project.

Management Measures

In response to local concerns, and in concert with state and federal regulations, the City of Burlington developed a source water protection program (SWPP), which was approved by the Vermont State Water Supply Division in 1994. Development of the SWPP was managed by the city, with assistance from the state, and enhanced by the involvement of businesses and concerned citizens. Public meetings were held to discuss the state SWPP development. During the development and implementation of the SWPP, the state met with local farmers to discuss strict agricultural standards for the management of spring runoff of nutrients from still-frozen fields onto which manure had been spread over the winter. The purpose was to work with the farmers to achieve an equitable balance between the water quality mandates of the state and the financial concerns of the agricultural community, instead of handing down unattainable standards. Some management measures adopted were a moratorium on manure spreading on still- frozen fields and/or building lagoons for holding waste until the frost leaves the ground. The SWPP is complemented by the additional protection efforts described below.

State of Vermont Act 250

In addition to the existing federal regulations, the State of Vermont has promulgated comprehensive laws through Act 250, which each town and city within the watershed must follow, regardless of the towns' individual differences in regulatory protocol. These regulations govern many of the land uses within the watershed, providing carefully monitored protection strategies that augment Burlington's protection efforts. The Act regulates the land use permitting process near lake shores and rivers, taking into account sewer system plans for new developments near Lake Champlain and its tributaries as well as timber management, proposed impervious surface areas, water use by ski areas to make artificial snow, and other issues. The Act helps to preserve the quality of the lake both within and outside of Burlington's jurisdiction.

City of Burlington Discharge Management Measures

Municipal wastewater treatment is commonly used to treat wastewater before discharging to surface water bodies. When Burlington's first treatment plant was built in 1951, a single pipe carried combined stormwater and wastewater flows. As the city grew, problems arose during storm events, when excess flows overwhelmed the system by 20 to 40 times its capacity, causing subsequent discharge of untreated sewage into Lake Champlain. Pollution events such as these piqued residents' concerns, leading the city to address the system's design. In the mid 1980s, several public forums were sponsored by the Department of Public Works to gain public input on these problems and generate consensus for solutions.

The result was a US$52 million bond vote for a system upgrade, which was completed in 1996. Due to the prohibitive costs of separating wastewater and stormwater systems, as well as the pollution caused from direct discharge of stormwater, the city chose to maintain a combined sewer system at its main plant and provided separation at its two smaller facilities. The main plant was refitted with a 10-foot diameter outfall pipe extending 2,500 feet into the lake, providing additional contact time for disinfection and allowing the wastewater to be diffused more easily in the deeper, more turbulent waters offshore. In addition, the plant's tertiary capacity was increased to 5.3 million gallons per day and a vortex separator was constructed to treat up to 75 million gallons per day of combined sewer flows during periods of rainfall or snowmelt. At Burlington's other two plants, combined flows were separated.

Watershed Protection and Management

The city's engineering staff have also been seeking solutions to the poor water quality of two tributaries that drain into Lake Champlain, the Winooski River and Englesby Brook. Englesby Brook provides drainage to approximately 600 acres of mixed use land in Burlington and South Burlington. It is subject to hydrologic, chemical and bacterial loadings typically associated with urban streams. Recently, the city received a grant to perform a watershed restoration project for Englesby Brook, which has been studied for the last ten years. The city started by teaming with the USGS to establish a stream gauging system and hired the Center for Watershed Protection (a nationally recognized nonprofit organization dedicated to the protection of watersheds), to develop a watershed management plan. Management measures addressed in the plan include the following:

- Retrofitting existing stormwater ponds for better flow control and pollutant removal
- Construction of new treatment systems
- Stabilization of the stream channel to prevent soil loss through erosion
- Development of a public outreach and education program.

The public outreach component of the plan in intended to help reduce pollutants generated at their source. Components of this program will include pet-waste management, lawn-care practices, roof-drain relocation, illicit connection detection/removal, increased street sweeping and dumpster waste management.

Drinking Water Intake Management

As water is withdrawn from the lake, it is screened and treated with chemicals such as chlorine or potassium permanganate to prevent colonization of zebra mussels inside the pipe. At the plant, the water is processed using coagulation, filtration and disinfection systems. Online turbidimeters measure the clarity of the water and particle counters measure the size and number of particles entering the distribution system. Burlington's water plant is a member of the Partnership for Safe Water, a proactive organization whose purpose is to encourage water suppliers to provide the best possible water treatment, focusing particularly on pathogen reduction.

US Case Study 3: Land Acquisition to Control Uses of the Watershed, Seattle Public Utilities, Seattle, Washington (USEPA 1999)

- Primary Sources of Water: Cedar River Watershed and the South Fork Tolt Watershed
- Watershed Area: 90,495 acres (Cedar River); 13,390 acres (South Fork Tolt)
- Population Served: 1.2 million
- Treatment: Chlorinated, unfiltered

To control the use of land within its watersheds, the Seattle Public Utilities (SPU) has made land acquisition a top priority in the Cedar River and South Fork Tolt watersheds, which are the primary sources of Seattle's drinking water. The utility owns 99.9 percent of the Cedar River Watershed and 70 percent of the South Fork Tolt Watershed. Land acquisition is funded with revenues from customers. SPU and the US Forest Service began exchanging land in 1962. This land exchange arrangement was formalized with passage of the Cedar River Watershed Act of 1992.

The law directs the US Forest Service to exchange critical Cedar River Watershed lands to SPU and places deed restrictions on the property prohibiting the City of Seattle from reselling, harvesting timber from, or developing (other than for routine maintenance) the parcels in question. These restrictions were instrumental to reducing the land's value, enabling SPU to afford the property through exchange of other parcels it owned worth approximately US$8 million. Without the restrictions, the land was valued at nearly US$100 million. This latest exchange brings the utility's total acquisition in the Cedar River Watershed over the last 100 years, to 99.9 percent of the land.

South Fork Watershed

The other area critical to Seattle's drinking water supply is the South Fork Watershed. Working with private industry, the utility completed a land exchange with the Weyerhauser Company for some SPU holdings in the North Fork Tolt basin. These properties had once been logged and are now in various stages of young second-growth forest.

This land exchange consolidates the holdings of SPU and Weyerhauser in the North and South Fork watersheds, thus improving the management of these lands. Although the utility's land is surrounded by Weyerhauser property, a signed agreement shows the support of Weyerhauser to protect water quality in the watershed. The city has now acquired 70 percent of the watershed lands.

The remaining eastern end of the South Fork Tolt is old-growth timber owned by the US Forest Service. No logging is taking place in this federally-owned area because there is an awareness of its value to the drinking water supply of Seattle.

US Case Study 4: Utility and Government Partnership, Syracuse Water Department, Syracuse, New York (USEPA 2005b)

- Primary Source of Water: Skaneateles Lake
- Watershed Area: 73 square miles
- Population Served: 160,000
- Treatment: Chlorination, fluoridation
- Partners: County Board of Health, Local Governments, New York State Department of Environmental Conservation

Skaneateles Lake, the fourth largest of the Finger Lakes in New York State, is long and narrow, averaging only three quarters of a mile in width over a length of sixteen miles. It is also deep, with a maximum depth of 300 feet. The total watershed area lies within three counties (Onondaga, Cayuga and Cortland), encompassing one village and seven towns. It is an oligotrophic lake, meaning that there are low concentrations of nitrogen and phosphorus, and relatively little productivity activity; consequently, the water is of high quality for drinking water purposes. Water quality monitoring has confirmed that nutrient and pollutant levels are not increasing in the lake.

The City of Syracuse, the Village of Skaneateles, and various water districts in the Town of Skaneateles and the Villages of Elbridge and Jordan all obtain water for their public water supplies from Skaneateles Lake via two 54-inch steel intake pipes belonging to the City of Syracuse (both located at the north end of the Lake). In addition, an estimated 1,000 lakefront residents have private intakes, although it is unknown how many of those households drink the water obtained through their own intakes. The lake is not used for industry, agriculture, or for animal watering, and the only known commercial use is for the irrigation of the Skaneateles Country Club. Boating and fishing are popular pastimes on Skaneateles Lake.

Given the current high quality of Skaneateles Lake water, the significant financial savings associated with a filtration waiver issued to the City of Syracuse and the Village of Skaneateles by the state (it is estimated that construction of a filtration plant would cost Syracuse US$60 millioni to US$70 million and the Village of Skaneateles US$4 million to US$6 million), and the high recreational value of the lake, the citizens of Syracuse and the Skaneateles Lake watershed have a vested interest in protecting their source of water from contamination. Efforts to protect Skaneateles Lake include the Skaneateles Lake Watershed Agricultural Program (SLWAP), the Skaneateles Lake Watershed Land Protection Program, several educational and outreach initiatives, and municipal regulations that govern zoning, proper use of chemicals, wastewater disposal, and agricultural activities.

Local Involvement and Developing the Protection Plan

Residents of the Skaneateles Lake watershed are involved in source water protection both through direct participation in protection efforts and as the targets of educational initiatives. SLWAP is a cooperative effort between the Cornell Cooperative Extension (CCE) Associations of the three counties in the watershed, several government agencies, the City of Syracuse, and farmers within the watershed. The program was initiated in 1994 as part of the filtration avoidance criteria established by the New York State Department of Health. The mission of the program is to implement Whole Farm Plans, based on Best Management Practices, to reduce the risk of nonpoint source pollution (pathogens, nutrients, pesticides, and sediments) from agricultural operations to preserve the water quality of Skaneateles Lake. As of 1997, 48 of 58 eligible farms were enrolled in the program, placing over 90 percent of the farmland in the watershed in SLWAP. By April, 2000, Whole Farm Plans were adopted by 26 farms and had resulted in a reduction in soil erosion of 2,700 tons per year, and farmers were saving an average US$1,000 per year in reduced fertilizer purchases.

The SLWLP program is another cooperative effort between the City of Syracuse Water Department and landowners within the watershed. The Finger Lakes Land Trust provided assistance in the early stages of the program by assisting with development of a draft easement plan and by conducting several "Talks and Treks" programs throughout the Skaneateles Lake area for public education.

The CCE offers public outreach tools to residents of the Skaneateles Lake Watershed as part of the national Home*A*Syst program, which is a pollution prevention and risk assessment program partially funded by the US Department of Agriculture and the US Environmental Protection Agency. The CCE has an advisory committee composed of residents of the watershed and representatives from local governments and nonprofit agencies. The CCE has a comprehensive website on its Home*A*Syst program (www.cce.cornell.edu/onondaga/homeasyst/index.htm) that links to information on pollutants, septic system management, water testing and treatment, well management, erosion control, and workshops. In addition, "Water Quality Educators" from the CCE are available to help residents assess pollution risks and provide information on how to lower the risks of contaminating water supplies.

Management Measures

In addition to the SLWAP and the Skaneateles Lake Watershed Land Protection Program, which are voluntary programs to manage potential point and nonpoint sources of pollution, there are several regulatory measures in place to further ensure the continued high quality of drinking water from Skaneateles Lake.

New York State Environmental Conservation Law 17-1709 prohibited point source discharges within the Skaneateles Lake watershed. Presently there are no municipal or package wastewater treatment plants or industrial discharges to Skaneateles Lake or its watercourses. Other potential sources of contamination include six stormwater outfalls from the Village of Skaneateles, a few smaller stormwater discharges, underground storage tanks, and nonpoint source pollution from septic tanks, agriculture, and erosion. Because nonpoint source pollution is perceived as the greatest threat to water quality, the Watershed Rules and Regulations, also implemented under New York State law, focus on evaluating these diffuse sources.

In 1996, both the Village and Town of Skaneateles updated their zoning and land use regulations, anticipating that the changes would offer improved protection of the public water supply. However, the other towns in the watershed (Niles, Spafford, Scott, and Sempronius) have few, if any, zoning measures, with the exception of lot size and setbacks in place for the protection of Skaneateles Lake water quality. Examples of the improvements in resource protection accomplished by the Town of Skaneateles' new zoning law include the creation of a Lake Watershed Overlay District, a Floodplain Overlay District, performance standards, size limitations, and Conservation Density Subdivisions. In 1993, the Village of Skaneateles implemented a cluster housing amendment to the zoning law to help preserve open space by requiring a conservation easement for cluster housing subdivisions. The two municipalities are also collaborating on a Joint Comprehensive Plan that will be used to guide future social and economic growth and development in the area. The ultimate goal is to maintain the quality of Skaneateles Lake water and groundwater at the current level.

The Skaneateles Lake Watershed Land Protection Program promotes water quality stewardship through the purchase of conservation easements offered voluntarily by landowners. Each easement, which limits development or logging, becomes a permanent part of the property title that must be adhered to by future owners of the land. In return, landowners can receive significant reductions in property taxes because the easement no longer is assessed as developable real estate. Usually, landowners donate conservation easements (the value of which are determined by an assessor based on the difference between the fair market values of the land with and without an easement), but the City of Syracuse has made limited funds available to the Land Protection Program that may be offered to eligible landowners.

Contingency Planning

In the event of a hazardous material or petrochemical spill, the towns within the Skaneateles Lake Watershed rely on three parties for emergency response and remediation. Local volunteer firemen are usually first on the scene and provide initial public and environmental protection. The county fire coordinator manages the local and regional response, coordinating the efforts of different fire departments based on their expertise or specialized equipment.

The New York State Department of Environmental Conservation Spills Response Team responds for containment of the spill and cleanup.

Measuring Program Effectiveness

The Onondaga County Soil and Water Conservation District designed a program to study the effects of agricultural BMPs on surface water quality in the Skaneateles Lake watershed. Parameters measured included phosphorus, nitrogen, coliform, and physical properties of the water (temperature, flow rate, etc.) Prior to implementation of BMPs, two dairy farms were chosen in the Grout Brook subwatershed to gather baseline data. Unfortunately, both farms went out of business, so the latest data will reflect the impact of removal of dairy herds rather than the effect of BMPs. The Otisco watershed, which is adjacent to the Skaneateles watershed, is conducting a similar study evaluating the effects of agricultural BMPs on water quality. Because the two watersheds have similar geology and hydrology, it is hoped that the results from the Otisco watershed will be transferable to the Skaneateles watershed. In addition, the City of Syracuse is conducting an evaluation of the SLWAP by interviewing farmers who have implemented Whole Farm Programs. The evaluation covers all aspects of the SLWAP by assessing farmers' experiences from planning to implementation and management of BMPs. Finally, water quality monitoring remains an effective tool for assessing drinking water protection measures.

US Case Study 5: Land Use Management, New York City and Seven Upstate New York Counties (USEPA 2005b)

Primary Sources of Water: Croton Watershed; Catskill and Delaware Watershed (includes 19 reservoirs)

- Watershed Area: 1,969 square miles
- Population Served: 9,000,000 (includes some upstate communities in addition to New York City)

The source water protection areas for this water supply are located in seven New York State counties: Delaware, Green, Schoharie, Sullivan, Ulster, Putnam, and Westchester. Agriculture is one of the major land uses in the source water protection areas. Dairy and livestock farming present one of the greatest non-point pollution challenges to the comprehensive source water protection program. The challenge regarding agriculture is to reconcile the public health and environmental resource protection interests of a large and distant city with the farming community's desire to maintain an agricultural way of life in the watershed region. Another major challenge is reconciling New York City's interests with those of upstate communities, including their interest in maintaining economic viability.

The major impetus for serious negotiations on a viable program to safeguard the city's water supply was provided by USEPA's clear intentions to require the city to filter its Catskill/Delaware water supply system (at the cost of several billion dollars) unless it substantially strengthened its existing watershed protection program. The New York State Governor's office and USEPA played a key role in getting the negotiations moving by bringing in all watershed stakeholders, including several environmental groups and a coalition of watershed towns. In 1997, the City of New York, New York State, the counties, watershed municipalities, and a number of environmental groups entered into a watershed protection agreement, called the Watershed Memorandum of Agreement (MOA), which is designed to protect New York's drinking water supply source for years to come. The partnership also includes the agricultural community and the federal government. The MOA consists of three separate watershed protection programs.

Land Acquisition Program

The Land Acquisition Program will enable the city to acquire environmentally sensitive, undeveloped land at fair market value from willing sellers. The city will continue to pay property taxes and conduct a community review process for any property under consideration. Because the parties recognize that any land acquisition program must provide reasonable opportunities for growth in and around existing population centers, the agreement allows some towns and villages to exempt certain areas from solicitation under the program.

The city is committed to spending up to US$310 million on land acquisition in the watersheds through outright purchase or conservation easement over a ten to fifteen year period. In October of 1997, the city acquired the first parcels of upstate land under the agreement. As of the end of 2000, the city had acquired or had purchase contracts to acquire over 26,000 acres from over 360 landowners at a total price of approximately US$73 million.

Watershed Regulatory Program

Under the agreement, new regulations were negotiated to control pollution within the watersheds. These regulations ensure that new projects are designed and constructed in ways that protect water quality within the watersheds. They include extensive review of proposed developments to ensure compliance with watershed regulations and standards.

Watershed Protection and Partnership Programs

In response to farmers' concerns about the potential economic impact of proposed revisions to New York City's watershed rules and regulations, the city put aside its purely regulatory approach and entered into a partnership with the watershed farm community to carry out a locally developed and administered voluntary Watershed Agricultural Program. The city committed a total of US$35.2 million to refine and demonstrate an environmentally sound whole farm planning approach. The farmer-led Watershed Agricultural Council has been very successful in recruiting farmers to participate in the program and in implementing whole farm plans. This is the one partnership program that is not in the MOA, having been put in place prior to MOA negotiations.

As part of its financial commitment to the MOA, the city pays for upgrades to the wastewater treatment systems owned and operated by upstate municipalities and private operators (e.g., the Wastewater Treatment Plant Upgrade Program). The MOA has also provided for nearly US$14 million in *good neighbor* payments, which go directly to the municipalities involved as payment for joining the agreement. Other major Watershed Protection and Partnership Program efforts described in the MOA include the Catskill Fund for the Future, septic tank rehabilitation, sewer extensions, new infrastructure, stormwater retrofits, stream corridor protection, sand/salt facilities, and public education (see more details on some of these efforts below). When the Watershed MOA took effect, the USEPA issued a five-year Filtration Avoidance Determination, which allows the city to avoid water filtration as long as management measures prove effective.

REFERENCES

Black PE. 1996. *Watershed Hydrology*, Second Edition, Ann Arbor Press, Chelsea, Mich.

Butcher JB. 1999. 'Practical Decision Methods for Watershed Management' *Human and Ecological Risk Assessment*, Vol. 5, No. 2, pp. 263–274.

Crowther R and Partners. 1999. 'Strategic plan for Water Management', Volume 1, *Water Supply and Demand Management*. www.crd.bc.ca.water/reports/strategic/vol3.htm

Field R, Tafuri AN, Muthukrishnan S, Acquisto BA, and Selvakumar A. 2005. *The Use of Best Management Practices (BMPs) in Urban Watersheds,* DEStech Publications, Inc. ISBN: 1-932078-46-0. 282 pp.

US Environmental Protection Agency. 1999. 'Protecting Sources of Drinking Water', *Selected Case Studies in Watershed Management,* USEPA 816-R-98-019, Office of Water, Washington, D.C.

USEPA. 2000. *Community Involvement in Drinking Water Source Assessments*. EPA-816-F-00-025. Office of Water, Washington, D.C.

USEPA. 2001. *National Menu of Best Management Practices for Storm Water Phase II*, cfpub.epa.gov/npdes/stormwater/menuofbmps/menu.cfm Washington, D.C.

USEPA. 2005a. *State and Federal Source Water Assessment and Protection,* EPA-816-R-05-001. Office of Water, Washington, D.C.

USEPA. 2005b. *Case Studies of Local Source Water Protection Programs*, www.epa.gov/safewater/protect/casesty/index.html.

Water Environment Federation and American Society of Civil Engineers. 1998. *Urban Runoff Quality Management,* 1-57278-039-8 and 0-7844-0174-8, Water Environment Federation, Alexandria, Va.

5 LAND USE PLANNING CONTROLS

Rob Walton, Robert Ford

CONTEXT

The quality and quantity of raw water available for any water supply system is dependent upon the activities performed in the watershed (catchment) area. For groundwater supplies, activities in the aquifer recharge area will also influence water quality, especially where shallow wells are used. Recent water quality incidents in Walkerton (Canada) and Sydney (Australia) both demonstrate the need for appropriate watershed protection, groundwater protection controls and general good management. In the case of Walkerton, Ontario, microbiological contamination of a shallow aquifer resulted in the deaths of seven people and thousands of illnesses. Although no health impact was identified in the Sydney incident, the resulting outrage and loss of community confidence in the water supplier cost Sydney Water over A$100m in compensation, testing and investigations; and caused the government of New South Wales to create a separate authority legislated to manage the watershed.

Both AWA and AWWA recognise the importance of multiple barriers for the protection of drinking water quality. The management of the watershed is the initial barrier to prevent contamination. Mechanical treatment systems are seen by some to overcome the need for watershed management, but they can fail or operate at lower than expected efficiency. Reduced treatment efficiency usually occurs when water quality is changing rapidly and the plant is heavily loaded with polluted water. Unfortunately it is usually at these times, during a storm event, when the highest level of microbiological contamination occurs, that the maximum efficiency is required to protect public health.

Any significant development of land or changes in land use in the watershed or aquifer recharge area can have the following impacts on the water supply:

1. Alter the quantity and quality of raw water available
2. Increase the level of water treatment needed to meet public health requirements and hence the cost (both capital and operational) of water treatment plants
3. Increase the difficulty of treatment by causing rapid changes in raw water quality, often by making the catchment runoff more vulnerable to storm events
4. Increase the health risk to be managed
5. Impact on the reliability of the water resource (by consumptive use or altering the watershed response to a storm. Usually increasing the immediate runoff, reducing infiltration to the water table and reducing base flow)
6. In the case of chemical contamination, may render the source unfit for consumption.

The governance of each particular jurisdiction will determine how planning decisions for land use and development are made. In most cases there is a hierarchy in the decision processes involving federal, state or provincial governments to set higher level policy decisions or policy frameworks. Setting local priorities for land use, zoning or the issuance of building permits is usually the function of the local municipal planning agency or commission. High level planning policy at a state or federal level is a critically important tool to ensure the quality of water sourced from local watersheds is considered when planning policies are implemented at local levels. Failure to communicate the importance of watershed management during the decision-making processes at any level can result in inadequate land-use controls or zoning or even loss of the ability of a water supplier to have later input.

Possibly the most important thing that any water supplier can do for land-use planning is to understand the hierarchy and decision points in the process. Once this is understood, it is imperative to ensure that the water supplier is recognised as a stakeholder in the decision making process and is well prepared to give good background information (i.e., science-based) to influence recommendations and, ultimately, decisions.

There are two levels of planning that need to be considered: strategic and local planning.

Strategic Planning

This involves basic decisions about the long term development and use of land at a regional and state level. At this level, water agencies have the potential to influence a number of important decisions.

The planning process should specifically require watersheds supplying drinking water to be identified on planning maps so that policy makers, local planners, developers and potential purchasers are all fully aware of the existence and importance of the watershed.

State/provincial environmental policies should identify drinking water sources as areas requiring special protection. In some jurisdictions this has led to a requirement for a higher standard for the design and monitoring of on-site wastewater disposal systems in watersheds, prohibition of inappropriate land uses (including higher density residential and industrial usage), prevented the inappropriate zoning of parts of watersheds and established a culture of precautionary decision making for watersheds.

Strong, strategically based planning support for local planners should be created, with a well developed policy framework. This is important as it makes it harder for less intensive (but still inappropriate) use and developments to be granted planning permission. Clearly, if water quality is not an issue in the strategic policy framework, it will be difficult to introduce water quality as an issue during the consideration of any specific land use or development application.

Local Planning

This involves local decisions about the granting of approvals for individual developments.

Rarely does the water supplier have full planning, or management, control over its watersheds (although in some jurisdictions it may have *veto* powers over some uses). The objectives of the local government agency are often different to that of the drinking water supplier (e.g., the promotion of economic development of the region is usually a high priority, exceeding the consideration of potential public health implications). In many situations the incremental nature of land development means that the danger to water quality is not recognised until large areas have been developed. By that time the situation is irreversible through land use planning. Consequently water suppliers need to anticipate source protection requirements and plan their involvement.

Important activities include:

1. Understanding the local planning governance and being *in the tent* for decision making
2. Demonstrating to the planning authority the need for watershed protection (social, environmental and economic)

by having completed the required background studies such as defining intake protection zones for surface watersheds and capture zone modelling for groundwater wellfields, compiling potential contaminant inventories and undertaking risk assessments. This is easier if watershed protection is a high level state/provincial policy
3. Demonstrating the value of a clean, low-cost water supply to the future development of the region
4. Influencing planning decisions to include consideration of the impact on water quality, by the development of well-researched water protection policies and subsequently insisting on consistent application of these policies to local planning decisions
5. Through consultation, and policy development, encouraging a mindset at the board level of water supply administrations (where policy framework is usually set) to oppose inappropriate development proposals that place water quality at risk. (This may result in political pressure being brought to bear from the development community, hence the need for good science and a strong conviction)
6. Making sure the watershed limits (surface watershed or groundwater capture zone) are identified on planning maps and work with the local municipal planning staff to develop appropriate land-use policies for these areas
7. Making senior management and water board/commission members aware of the need to protect the watershed so they can use their influence to direct strategic planning decisions
8. Ensuring relevant stakeholders (planners, management, and directors) properly understand and evaluate the impacts of land-use development on water quality and quantity (by providing a sound technical basis for land-use policy and decisions)
9. Engaging the public in relevant debate about the effects of development on source water quality and quantity
10. Educating staff about the need for watershed protection so they are able to support the process within the community
11. Understanding that some problems are common to most water suppliers and develop local ways to deal with them. These problems include:
    * Powerful lobby groups such as developers, agriculture and business often have good connections to the politicians for decision making
    * Political terms are much shorter than the planning horizon for water supply, which can lead to decisions that are not geared to sustainable development or, simply, to rapid turnover of decision makers that the water supplier needs to be aware of
    * Legal and regulatory systems do not always match with applying risk management.

OPTIONS

While watersheds that are closed to human activity are ideal for source water protection, this is an impossible option for most water suppliers. Accordingly, water suppliers need to evaluate their options to maintain or develop sustainable water supplies.

Land-use planning ordinances provide an opportunity to:

1. Place the existence of the watershed in the public domain through planning policies and maps
2. Give legal status to guidelines/standards required for development in the watershed by incorporating them into the planning system
3. Develop clear, transparent processes and policies for assessing development applications
4. Obtain water quality improvements by negotiation with developers and landowners
5. Obtain improved water quality controls during the construction phase of a development
6. Lock in requirements for long-term maintenance of controls (this can be a

difficult mechanism to monitor in some jurisdictions)

7. Generate support from environmental interest groups.

Types of control that can be considered in an ordinance include:

1. Defining permitted land use activity in various parts of the surface watershed or groundwater capture zones and limiting higher risk development to appropriate areas remote from watercourses or outside of the vulnerable areas of a capture zone or intake protection zone. This is often done on a risk assessment/sliding scale basis as outlined in the Case Studies portion of this chapter
2. Prohibition of inappropriate land uses (industries/developments) that are considered to be a very high pollution risk, for example, a landfill or land uses that attract large numbers of people/animals into the watershed or capture zone
3. Controls on density of development
4. Fencing and revegetation of streams and waterways to prevent livestock access and creation and maintenance of vegetated stream banks as a condition of granting planning consent
5. Management of unsealed roads/verges to control erosion (especially tracks for fire control or logging)
6. Management of agricultural practices such as nutrient application (in some jurisdictions this is not available within the planning process) as part of the permitting system for new agricultural structures or changes in farming practices, usually resulting from livestock intensification
7. Delineating setback distances from waterways and reservoirs for dwellings and on-site domestic wastewater treatment and disposal systems. Require the provision of reticulated wastewater collection systems and stormwater treatment systems as a condition of permitting further development. Management and audits of on-site water treatment systems throughout the watershed as part of the approval system
8. Training front counter staff at municipal offices to identify applications for potential high-risk development at the earliest possible stage and referring the applicant to the water supplier
9. Developing a protocol with the planning office whereby it circulates all planning applications at the earliest possible date to the water supplier that pertain to lands within delineated surface water intake protection zones or groundwater capture zones, as part of the application decision process
10. Purchasing land and/or placing restrictive controls over sensitive areas
11. Requiring formal agreements, registered on the property titles, specifying the management of a parcel of land in a certain way (this can be in consideration of approving development of the site). In some jurisdictions this option may not be useful, depending on the jurisdictional laws.

It is important to know the legal ramifications and justifications for the measures proposed and to check with the planning office to see which controls can be achieved within the legislative authority of the municipality. These will vary between jurisdictions.

BEST PRACTICE

This section sets out current suggested best practices that water providers from around the world have used to promote sustainable development within their jurisdictions. It is essential to become active in strategic planning in the region. To be effective, an organization needs a clear vision of where it needs to be going and what it hopes to achieve in the planning documents. An in-house champion needs to be appointed and supported to manage watershed issues, or else ensure that agency structures clearly define responsibility for watershed management activities.

The best practices listed are intended to assist water suppliers with the types of comments/input that they should have on planning applications. The list may not be exhaustive but is intended to provoke thought and hopefully innovation for use in any particular jurisdiction.

Commenting on Planning Policies or Applications

The best practices listed below have been implemented by water suppliers involved in proposed planning policies or planning applications:

- Zoning the bulk of the catchment for the activity that results in minimum impact. This is often (but not always) a broadacre agricultural use that limits the amount of human or animal activity. (Most farmers have an affinity with the land and understand the need for erosion control and care in the use of agricultural chemicals). In some cases large rural/residential lots may be appropriate with strong controls on animal keeping, vegetation protection and enhancement and waste management
- Allowing farming to continue as a permitted use with stewardship responsibilities to prevent pollution (i.e. nutrient management planning)
- Opposing additional dwellings where the density exceeds a specified limit on private services (water and septic systems). One method of achieving this is to consider the number of dwellings in, say, a 1 mile or 1.5 km radius and measure/model pollution levels. From this data guidelines with appropriate factors of safety can be set
- Ensuring all urban and semi-urban development have sewerage or a strong management plan for on-site disposal systems and appropriate treatment of storm runoff
- Requiring an appropriate setback from any watercourse for all new dwellings and domestic wastewater treatment systems. (Could be reduced in appropriate locations if there is a system in place, managed by a statutory agency to the satisfaction of the water supplier, that ensures domestic wastewater plants are audited and serviced regularly and defects are rectified promptly)
- Only approving subdivision of land where it is suitable for appropriate agricultural purposes or the water supplier is satisfied that wastewater and storm runoff can be treated and managed on all new allotments
- Coming to an arrangement with the planning agency that all development applications in the watershed will be circulated to the watershed agency. Strengthen this protocol by having the planning department directly discuss any applications that fail to meet agreed guidelines
- Formally opposing inappropriate development by seeking an independent review
- Ensuring that all significant decisions, policies and the reasons for them are recorded in the public domain and that all water suppliers' comments on planning applications are in writing.

Research and Other Management Practices

Modeling of the impact of development can be easily accomplished using pollution export modelling. This can feed into a risk management process and be used to set guidelines. Better data to verify model parameters under different climatic conditions and land usage are needed.

For groundwater watersheds, three-dimensional models of the capture zone can be used in risk assessment for future land use designation. As outlined in the case studies, many jurisdictions have used risk assessment to categorize where different land uses will be prohibited, permitted or permitted with certain controls in place.

One area of concern to all water suppliers is existing uses that are not located appropriately for source water protection. In most jurisdictions these uses have the right to exist and continue. Another major risk for water supplies is abandoned properties such as old industrial sites and old gas/petrol stations. A best practice that many municipalities have implemented is an incentive program to help pay for facility upgrades that promote water protection. These programs usually involve facility inspections, evaluations of risk areas and proposals for upgrades to mitigate the risks. Funding for such programs may come from the municipal tax base or water rates and is sometimes subsidized by senior levels of government.

Conducting internal research to prove something felt to be important is an option where no answer is available off the shelf. Universities are often very interested in such projects and may have matching funding sources. For example, the transport of nitrates from agricultural operations to a groundwater aquifer depends on the soils, application rates of manure or fertilizer, rainfall rates, drainage, etc. In a particular watershed, where this is deemed a threat, specific research may be required to determine what level of livestock intensification would be a threat to the groundwater supply.

CASE STUDIES

County of Oxford — Wellhead Protection Pilot Study

The County of Oxford is a rural municipality in Ontario, Canada. The total population is 100,000 and the land area is 2,003,000 hectares. The water supply in this area is 100 percent groundwater. The County has planning authority and is the local water supplier for 70,000 people (80 municipal wells for 19 water systems). The rural population gets its drinking water from approximately 11,000 privately owned groundwater wells.

Owing to concerns raised over the development of large livestock operations and the resulting manure production, the County undertook a Groundwater Protection Study in two phases from 1998 to 2001. Three-dimensional modelling of the municipal well capture zones was undertaken in Phase 2. The total cost of the studies was approximately C$850,000. Approximately 10 percent of the total land area of the County is in municipal well capture zones and approximately 40 percent of the County is in a highly vulnerable aquifer recharge area. Phase 2 of the study had numerous recommendations including amendment of the county official plan to strengthen groundwater protection policies.

When the County was in the process of amending the official plan to include land use controls in municipal well capture zones, two area municipalities raised concerns about the potential effect on economic development.

With the assistance of a provincial grant, the County and the two area municipalities, the City of Woodstock and the Town of Ingersoll, undertook a pilot study to look at how to implement source water protection measures in proposed development areas. The study had four components:

1. Legal review of the possible tools available through existing legislation and the legal ramifications of the measures proposed through the study
2. Hydrogeologic review of the capture zone modelling undertaken in the Groundwater Protection Study, Phase 2, including site drilling to calibrate model parameters
3. A risk assessment evaluation of possible land uses in the municipal well capture zone
4. Stakeholder involvement integrated into the pilot study.

The main outcome of the pilot study relating to land use planning was a list of prohibited uses and a list of conditionally permitted uses in the capture zone. The conditionally permitted uses will evaluate their potential groundwater risks using a checklist prepared as part of the study in order to determine best management practices, site development requirements and monitoring requirements (if any). This checklist will be administered at the building permit stage, or where a development approval is required, during the planning application process. A different form of the checklist may be implemented for existing businesses along with an incentive program to improve water protection. The pilot study can be viewed at www.county.oxford.on.ca/cao/planning/water_policy.asp.

This case emphasises:

- Local initiative to tackle issues that could threaten water supplies
- The importance of good technical studies to support planning decisions.

Oak Ridges Moraine

The Regions of York, Durham and Peel, situated east, north and west of Toronto in Ontario, Canada, take in most of the Oak Ridges Moraine. The Oak Ridges Moraine Conservation Plan (2002) was enabled through specific legislation, the 'Oak Ridges Conservation Act' passed in 2001. It guides municipalities in the development of land use policy to protect environmentally sensitive areas as well as ground and surface water. This plan is established by provincial regulation and provides land-use and resource management direction for the protection of the land and water of the Moraine.

Municipal planning decisions must conform to the Oak Ridges Moraine Conservation Plan, which takes precedence over their plans. Municipalities within the Oak Ridges Moraine must bring their official plans and planning documents into conformity with this plan, and are currently undertaking measures to do so.

The Oak Ridges Moraine Conservation Plan prescribes specific prohibitions and restrictions on the use of land within wellhead protection areas and vulnerable areas.

Subsection 28(1) of the Oak Ridges Moraine Conservation Plan requires that the following uses be *prohibited* in the wellhead protection areas:

- Storage, except by an individual for personal or family use, of:
  * Petroleum fuels
  * Petroleum solvents and chlorinated solvents
  * Pesticides, herbicides and fungicides
  * Construction equipment
  * Inorganic fertilizers
  * Road salt
  * Contaminants listed in Schedule 3 (Severely Toxic Contaminants) to Regulation 347 of the Revised Regulations of Ontario, 1990
- Generation and storage of hazardous waste or liquid industrial waste
- Waste disposal sites and facilities, organic soil conditioning sites and snow storage and disposal facilities.

Additional prohibitions apply in the 0–2 year time-of-travel zone. Some of these uses are also prohibited in highly vulnerable aquifer areas.

The Oak Ridges Moraine Conservation Plan directs municipalities to establish wellhead protection policies according to time-of-travel zones. The aquifer vulnerability analysis resulted in the mapping of two aquifer vulnerability classifications for the Oak Ridges Moraine: high and low. Certain of the above-listed uses are also prohibited in highly vulnerable aquifer areas.

The conservation plan also establishes that *existing* uses within the wellhead protection areas and highly vulnerable aquifer areas are subject to certain restrictions and additional site management and reporting requirements. This requirement is unique to the Oak Ridges Moraine and requires special enabling legislation, as these powers are not available under the Ontario Planning Act.

This case emphasises the importance of high level planning to coordinate source water protection in multi-jurisdictional areas.

Blackwood

This is a small unsewered community northwest of Melbourne, Victoria, in Australia. In 2002 the local planning authority erroneously permitted the construction of a dwelling on steep land adjacent to a stream, 600 metres upstream of the town water supply intake. The responsible water authority had a clear policy that had been adopted following public consultation that, generally, no dwelling or wastewater disposal system should be permitted within 100 metres of any waterway within the watershed.

The state government had issued strategic guidelines to local planning authorities recommending similar setbacks. The Victoria Environmental Protection Authority had also issued a code of practice for on site wastewater treatment that also recommended a setback of 100 metres where no formal monitoring program existed. The local planning controls required referral of all applications within 100 metres of a waterway to the water authority. However, because the planning agency had not identified the need for local mapping to alert its staff, this application was not referred and was granted approval.

The water authority sought enforcement of its referral powers and requested a review of the merits of the application at the appropriate legal jurisdiction. Because of the strong strategic support at a state policy level for watershed protection and the existence of a well documented local strategy, the review board ordered the dwelling to be removed.

This case emphasizes:

- The importance of high level planning strategy in supporting local decisions
- The value of well thought out and documented local policy
- The need for better communication with planning agencies.

REGULATIONS

The regulatory context for planning in regard to source water protection varies quite widely.

Canada

Water issues are generally handled by the provincial or territorial governments, except for international boundary waters and water protection initiatives on lands and waters of federal jurisdiction, including Aboriginal reserves. Federal government support of source water protection initiatives is considered to be very important to ensure intergovernmental coordination. Constitutional jurisdiction regarding source water protection is not always clear. In Ontario alone, the Federal government owns or controls 460,000 hectares of lands and waterways, not including Reserve lands as defined under the Indian Act and, therefore, its participation is important to the success of a source water protection program. The Provincial/Territorial regulations vary.

Ontario

Ontario has a Safe Drinking Water Act that provides a source to tap framework for drinking water. Province-wide legislation under the Act was proposed for source water protection in the fall of 2005. Source Water Protection Committees and Boards, based primarily on surface water watershed areas, or locally known as Conservation Authorities, will develop source protection plans based on the criteria set out in legislation and regulation. This high level plan will set out the criteria for the area municipalities in the watershed to follow in their planning documents and other actions required. Monitoring of source water protection activities may be undertaken on a watershed basis and by the local water supply authority. Local planning and economic development are undertaken by the area municipality. Local planning documents will be required to have regard for the source protection plan. The Province is funding the development of the source water protection plans, beginning in 2005.

New Brunswick

New Brunswick has a Provincial Clean Water Act which is an umbrella Act for several regulations/orders. Source water protection for surface water is undertaken through the Protected Area Designation Order, which protects surface water used for drinking water purposes by establishing a 75-metre setback and controlling development within these areas. For groundwater, source water protection is through the Wellfield Protected Area Designation Order, which divides the area around wellheads into three protection zones and cites specific uses and activities that are permitted in each zone as well as those that are permitted only under certain conditions. Permitted activities are prohibited from releasing contaminants onto the ground or into the aquifer or adversely affecting the quantity or quality of water. For private wells, the province has a Water Well Protection Program. The Province is responsible for classifying all watersheds. In New Brunswick, the cost of studies to support designation are generally shared between the Province and the local municipality.

Australia

In most states and territories, land planning is managed at the local municipal level; however, water supply varies from being a local government function in some states, to separate state-owned enterprises in others or even, in three jurisdictions, a state agency.

In all cases, however, state and territory governments develop high-level strategic planning policies. These usually include statements on the importance of watersheds, but this needs strengthening. Watershed authorities charged with coordinating and developing strategies for improving river health are being created in many areas; a total of 56 in mid-2007. Because of their strategic importance and ability to direct state and federal funds, water supply authorities are becoming increasingly involved in their activities.

The State of Victoria has enacted a Safe Drinking Water Act (2003) that requires all water supply authorities to develop and implement an audited risk management process extending from the watershed to the customers' taps.

United States

The Safe Drinking Water Act (SDWA) was passed in 1974 and amended in 1986 and again in 1996. The SDWA requires all public water systems to undertake a Source Water Assessment Program (SWAP). Each state develops its own SWAP for approval by the USEPA.

There is a lot of flexibility in how each state implements its program, but each program must include:

- Delineating (or mapping) the source water assessment area
- Conducting an inventory of potential sources of contamination in the delineated area
- Determining the susceptibility of the water supply to the contaminant sources
- Releasing the results to the public.

Communities can use different source water protection methods such as:

- Regulations to prohibit or restrict land uses that may release contaminants in critical source water areas
- Organising local events, distributing information and educating people/businesses to participate in environmentally responsible practices
- Purchasing land or creation of conservation easements for protection zones near drinking water sources.

The SDWA also included the Wellhead Protection Program (WHPP) to protect groundwater. The SDWA specified that program activities such as delineation, contaminant source inventory and source management be incorporated into state WHPPs which are approved by the USEPA before implementation.

WHAT SHOULD WE KNOW?

When challenged by those who question limits on economic development it is sometimes difficult to provide black and white answers, so it would be good to have more research. The following is indicative, but is not necessarily a complete list of research needs for land use planning effects on source water protection:

- Better benchmarking or calibration of watershed models to real life monitoring
- More documentation and benchmarking of cost-benefits of source water protection vs. economic development; sustainable development on a regional

or watershed area has to be considered, not small-scale modelling
- More development of programs/financing models/cost-benefit analysis of dealing with existing abandoned, polluted sites
- Documentation to justify land use restrictions to show:
  * Costs to society
  * Winners and losers
  * Compensation and cost recovery
  * Needs of many versus the gain of few
- Watershed models for surface water or capture zone modelling for groundwater have the best accuracy on a catchment or regional scale. Applications for small plot-scale planning have less accuracy and cannot predict long term effects of the proposed change. Research to calibrate these models to evaluate planning proposals more accurately on a smaller scale would be beneficial to water suppliers. It would also assist in assessing risk of existing uses.

OTHER RESOURCES

Useful Policy Documents

The Walkerton water tragedy was caused by complete failure of the *multiple barrier system*. With respect to the planning of the groundwater well locations, the placement and construction of Walkerton well #5 was instrumental as one of the multiple barrier failures in Walkerton. An Inquiry was undertaken to look at the causes of the tragedy and how to prevent a repeat. For detailed information on the Walkerton tragedy water professionals should read the Report of the Walkerton Inquiry, Parts One and Two by the Honourable Dennis O'Connor, Commissioner. Part One deals with the tragedy and what caused it.
Part Two makes recommendations to the Province of Ontario to strengthen its water supply legislation. The Province of Ontario has committed to implement all 121 recommendations of the two reports.

REFERENCES

Artemesia Waters Ltd. et. al. v. County of Grey (No. 2) (PL001055) July 24, 2001, Ontario Municipal Board Reports 42, p. 238.

County of Oxford Groundwater Navigator: maps.county.oxford.on.ca/groundwater/index.cfm?MenuItemID=4.

Environment Canada: www.ec.gc.ca/water/en/manage/qual/e_multi.htm.

Gold Mountain Springs Inc. v. Township of Oro-Medonte (PL000560) February 14, 2002, Ontario Municipal Board Reports 43, p. 129. www.omb.gov.on.ca/e%2Ddecisions/pl000560%5F%230287.pdf.

Hrudy, SE & Hrudy, JH. 2004. *Safe Drinking Water — Lessons from Recent Outbreaks in Affluent Nations*. pp 81–118 & pp 351–367. IWA Publishing, London.

Joint Catchment Policy of Four Regional Water Authorities: www.chw.net.au/policy.html.

Ministry of Municipal Affairs and Housing and B. Gardiner v. Township of West Perth (PL000064) re: Township of West Perth Zoning By-Law 100-1998; July 18, 2000, Ontario Municipal Board Reports 40, p. 428.

McQ Handling Inc. v. Town of Ingersoll (PL000793) November 21, 2002, Ontario Municipal Board Reports 44, p. 429 www.omb.gov.on.ca/e%2Ddecisions/pl000793%5F%231599.pdf.

O'Connor, D. 2002. *Report of the Walkerton Inquiry, Part One, The Events of May 2000 and Related Issues*, Queen's Printer for Ontario.

O'Connor, D. 2002. *Report of the Walkerton Inquiry, Part Two, A Strategy for Safe Drinking Water*, Queen's Printer for Ontario.

6 SURVEILLANCE

Clairly Lance, Jerry Schulte

CONTEXT

The primary objective of surveillance is the protection of water quality, i.e., to prevent pollution, and it requires the observation of activities and conditions in a watershed. From an Australian perspective, surveillance contributes to Elements 2–6 of the *Framework for the Management of Drinking Water Quality* (NHMRC & NRMMC 2004 and can be found on the NHMRC website at www.nhmrc.gov.au).

Surveillance programs have multiple values in relation to risk and vulnerability assessment. When establishing surveillance programs for a watershed a risk and vulnerability assessment should be used to determine the level of surveillance required; generally the higher the risks the more surveillance activities are required. Additionally, once established, the surveillance program feeds essential information back into the risk and vulnerability assessment process in the important area of *knowing your watershed* as well as providing both qualitative and quantitative information.

Surveillance has several objectives:

- Observation of general activities, events or incidents for acceptance within legislative requirements
- To meet the duty of care for a water utility to protect and understand its watershed
- Checking particular activities, events or situations for compliance according to specific standards or prescribed sets of conditions
- Validation and verification of risk
- Looking for anything unusual or unapproved
- Inspecting post-incident conditions (e.g., after a heavy rainfall event or wildfire)
- Proactive management of watersheds and being able to read the signs that warn of things going wrong; to have measures in place beforehand that reduce adverse impacts.

Surveillance is a complementary activity to monitoring, but has its own set of drivers and techniques, so has been addressed separately in this book.

OPTIONS

Options for watershed surveillance include regular, programmed, proactive surveillance and reactive surveillance. Proactive surveillance is the baseline surveillance required for a watershed and includes the inspection of stream, river or aquifer condition and of the watershed landform for signs of degradation, or possible contamination of water bodies because of:

- Land uses
- Natural causes, e.g., rain and storms
- Illegal activities, e.g., bylaw infringements
- Recreational activities

- Feral animals
- New land development
- Commercial land uses such as logging and mining.

Reactive surveillance occurs as a result of an event in the watershed. Events that can trigger reactive surveillance include:

- Wildfires, heavy rains and the like
- Pollution incidents such as chemical spills
- The need to prosecute or deter people doing the wrong thing or breaking laws in the watershed. Proactive surveillance may highlight evidence of such things as illegal fishing and reactive surveillance may be planned in an attempt to catch the offenders in the act.

In addition to traditional surveillance (i.e., officers driving around the watershed), other methods are available. These include aerial photography, which can be very valuable and, if available, should be linked to GIS and a database for information. Satellite photos or aerial surveillance can also add value to overall surveillance actions for a watershed, especially in areas of land management change; or to assess the effects of wildfire; and may also be useful for large, inaccessible areas.

Surveillance Frequency

The frequency of surveillance activities is closely linked to risk. Generally speaking, as risk increases the frequency of surveillance should increase. Ask questions to determine the factors that will influence the design of a surveillance program for a watershed:

- What type is the source (surface/groundwater) and the related watershed?
- If the source is groundwater, is it a confined or an unconfined aquifer?
- How close is the watershed to a population centre?
- What land uses are occurring in the watershed, and what are the associated water quality risks including past anectdotal evidence?
- What is the purpose of the inspection program? Is it to measure compiliance with set conditions, is it to educate land users, or is it both?
- What is the value placed on having a presence in a watershed?

Table 6.1 illustrates how surveillance activities can vary in different watersheds, based on land uses, risks and the proximity of the watershed to population centres. This table is a simplified model of how surveillance frequencies vary between different watershed types. It is important to note that one size does not fit all and watershed managers will need to design surveillance programs that meet their objectives for managing their specific watersheds. Table 6.2 details some key items to observe, points to note and follow-up action(s) required for best practice watershed inspection and surveillance programs.

BEST PRACTICE

One of the outcomes of conducting surveillance is finding pollution within a watershed. It is essential that plans/procedures and response protocols for dealing with pollution are in place before pollution is found. Response protocols may have a staged response with the level of response increasing with the level of contaminant found.

In addition to watershed officers collecting information such as the observations in Table 6.2, it is imperative that this information is passed on. There appears to be an underlying common problem for water utilities and watershed managers across both North America and Australia with on-ground watershed officers collecting invaluable information and not communicating effectively with their line managers, which results in a substantial amount of valuable information being left in field logs.

In order to solve this problem watershed managers need to provide their staff with a system to systematically record their observations from the watershed and, where required, escalate issues to an appropriate level of management. This system should allow a historical record of

Table 6.1 Baseline surveillance frequency for different watersheds

Watershed Description	Surveillance Requirements
Surface water source — pristine close to population centre • Land use: Pristine watershed with native forest/vegetation some timber harvesting/logging occurring • Activities: Some illegal activities including fishing and swimming especially over the summer months • Location: On the outskirts of a large regional town/city	• Minimum baseline surveillance is 0.5 – 1 day/week driving around the watershed, checking preventive measures/barriers and reporting on items listed in Table 6.2 • Timber harvesting /logging activities — weekly inspections when these activities are occurring, when harvesting completed continue to check on status of timber harvesting tracks and drainage, especially after initial rains for the wet season and significant rainfall events • Additional reactive surveillance activities occur as required post incidents (heavy rainfall, wildfire, pollution incident, etc.)
Surface water source — pristine remote location • Land use: Pristine watershed with native forest/vegetation • Activities: Small amount of illegal activities including fishing and swimming, especially over the summer months • Location: Water supply for a town of 50 people, located on the outskirts of town	• Minimum surveillance is 0.5 – 1 day/month driving around the watershed checking preventive measures/barriers and reporting on items listed in Table 6.2 • Other visits may occur to the watershed by staff undertaking raw water monitoring; these staff would report any obvious signs of pollution/illegal activities • Additional reactive surveillance activities occur as required post incidents (heavy rainfall, wildfire, pollution incident, etc.)
Surface water source — partly developed/fully developed • Land use: Small percentage of the watershed contains native vegetation, agricultural, industrial and recreational activities occur within the watershed • Activities: Industrial, agricultural and recreational activities occurring • Location: Water supply for a city/ large town located within the watershed	• Minimum baseline surveillance is 3 – 5 days/week driving around the watershed, checking preventive measures/barriers and reporting on items listed in Table 6.2 • Liaising with landowners/industries/recreational groups in the watershed on relevant best management practice techniques and applicable legislation • Depending on level of development/land uses, inspections may occur to check compliance with agreed conditions • Additional reactive surveillance activities occur as required post incidents (heavy rainfall, wildfire, pollution incident, etc.)
Groundwater source — unconfined aquifer with fenced bore compound and sealed bores • Land use: Developed recharge area with agricultural activities	• Minimum baseline surveillance is 0.5 day/month to liaise with landowners in the recharge area on relevant best management practice • Note: Resources allocated to surveillance will depend on size of the recharge area and land uses occurring within it. If recharge area is large with many land users, may choose to focus activities on land uses posing the highest risk to water quality (i.e., based on the type of industry and its location within the recharge area).
Groundwater source — confined aquifer with fenced bore compound and sealed bores • Recharge area: Located some distance from borefield, hydrogeology indicates that it takes thousands of years for water to reach bores from recharge area. • Land use: Recharge area is low intensity livestock grazing.	• Minimum baseline surveillance is 0.5 days/half year to inspect area around borehead and liaise with any land users in close proximity to the bore • Watching brief kept over recharge area especially for any heavy industry development, checking preventive measures/ barriers and reporting on items listed in Table 6.2

past incidents (even minor ones) and allow improved follow-up of unusual sightings (e.g., a cut fence led to a stolen truck being submerged in a reservoir, however the cut fence was not initially reported and was simply repaired).

CASE STUDY

Hypothetical surface water source with a 2 gigalitre capacity — unprotected watershed — state forest with recreation occurring throughout the watershed (hiking, camping, offroad vehicle driving), fishing swimming and boating occur on the reservoir body.

Inspection/Surveillance Program

- Surveillance of the watershed at least 5 days/week during recreational season specifically targeting areas where the public recreate as well as inspecting the watershed for any of the items listed in Table 6.2. Use opportunities to educate the public on watershed issues. Specific surveillance required at swimming sites; this will depend on local council/county requirements relating to swimmer safety.
- Communication network required with all recreational groups using the watershed, this could be in the form of a water utility representative meeting with recreational group committee members on a regular basis.

WHAT SHOULD WE KNOW?

How can appropriate frequencies for surveillance programs be rationally determined?

Are there any computer-based packages for source protection operations data (from surveillance) with linkages to drinking water quality programs?

Surveillances priorities and objectives — operational versus education, i.e., officers in the field can have an educative role as well as a policing one. Can that be effectively integrated with surveillance? Can it be justified on budgeting grounds?

Table 6.2 Items to observe during inspection and surveillance programs

Items to Observe	Points to Note	Follow-up Action
Access roads	Condition	Arrange remedial action if necessary.
Signage	Inspect and ensure the performance of signage in relation to the effect on visitor behaviour. Identify damaged signs and replacement trends/details	Post signage required to increase effectiveness (e.g., restrict access) and replace as necessary
Erosion	Location and severity	Arrange remedial action: short term — erosion control measures; long term — rehabilitation
Declared weeds	Type, location and severity	Arrange for eradication
Feral animal activity	Type, location and severity	Arrange an eradication program
Evidence of illegal activities (e.g., entry to restricted areas, swimming, fishing)	Location and severity	Arrange a specific surveillance patrol to target illegal activity
Dumping rubbish/chemicals	Type, location and quantity	Arrange removal, advise other agencies
New land development including rate of growth	Location, impact on water quality, compatibility with planning controls	If not compatible advise planning agency, appeal development approval, check for compliance with any special conditions required as a condition of the approval
Livestock activity	Location especially in relation to streams, reservoirs, well heads	Fencing and revegetation of streams and waterways to prevent stock access and creation of vegetated stream banks
Activities by other agencies (e.g., Forestry Department, local shire or county council)	Assess impact, monitor compliance with agreed standards	Arrange remedial action where required and where applicable arrange approvals process for future activities
Hazardous sites	Location, type, assess hazard	Arrange protective measures
Approved recreational events	Attend event and ensure compliance with event approval conditions, assess impact	Assess impact and arrange remedial action if required
Post wildfire	Inspect the burnt area and fire trails used, upgraded and installed. Inspect and map fire trails installed/upgraded. Where available obtain fire intensity maps from fire authority. Liaise/advise on the rehabilitation required from the fire.	Investigate potential for erosion problems resulting from the fire and associated activities. Monitor turbidity in problem areas, sample for nutrients in first heavy rains after the fire to monitor nutrient release. Arrange restoration/rehabilitation as required. Install erosion control measures on fire trails such as drains, booms, and sediment traps

7 MONITORING

Clairly Lance, Jerry Schulte, Pat Feehan

CONTEXT

The primary objective of monitoring is the protection of water quality, i.e., to prevent pollution. Monitoring techniques provide data to understand how activities/land uses/events that occur in a watershed impact drinking water quality and provide essential information, especially in relation to hydrological characteristics. Information from monitoring enables water utilities to manage their watersheds to optimise water quality, minimise risks to human health and protect the environment. From an Australian perspective monitoring programs contribute to Elements 2–6 of the *Framework for the Management of Drinking Water Quality* (which can be found on the NHMRC website at www.nhmrc.gov.au).

Monitoring programs have multiple values in relation to risk/vulnerability assessments. When establishing a monitoring program for a watershed, a risk/vulnerability assessment should be used to determine the level of monitoring required; generally the higher the risks, the more monitoring is needed. Additionally, once established, the monitoring programs feed essential information back into the risk/vulnerability assessment process.

Monitoring has four broad objectives:

1. Reviewing watershed management actions to determine if desired outcomes are being achieved and what changes are needed
2. Ensuring the effectiveness of barriers and watershed preventive measures
3. Monitoring water quality and land use activities and changes in an attempt to link activities with effects
4. Independent auditing of compliance with accepted watershed management guidelines and policies.

The specific objectives of monitoring programs include:

- Providing a baseline record (including peaks and troughs) of watershed data for individual sources
- Measuring seasonal and natural trends in water quality to provide information on the impact of changes in land use and of storms and droughts, and to detect contaminants in raw water that could impact on the water treatment process
- Measuring the impacts of different land uses on water quality, including understanding the contribution of various activities to water quality issues
- Measuring the impact of land use changes on the watershed, however incremental, over time and providing data to assist developing land use planning controls

- Achieving early detection of possible contaminants in the water and protecting the health of customers
- Identifying trends that may indicate possible sources of contamination (hotspots) in the watershed and determining the effectiveness of the watershed protection program (including preventive measures and barriers)
- Assisting watershed managers to identify emerging/changing watershed conditions (including water quality issues) so that they can be addressed in forward planning
- Reporting on performance against water quality criteria at different locations throughout the watershed
- Confirming results of watershed improvement actions
- Providing information on the impact to the watershed of incidents such as natural disasters and pollution spills.

It is important to recognise that watershed managers do not have endless budgets to resource monitoring programs. This is where a risk based approach to managing a watershed can be useful, with the major part of monitoring resources focussing on those activities that pose the greatest risk to water quality.

OPTIONS

Monitoring by itself does nothing to protect water quality; its purpose is to support the watershed protection activities. Monitoring can include water quality monitoring as well as recording and analysing land-use activities and the impact of individual events (e.g., spills, wildfire).

Monitoring includes baseline monitoring programs, which provide information on the normal or background characteristics for a watershed, and event based monitoring programs, which can provide data on hazardous events (once the base level value of parameters has been established the baseline monitoring program may be reduced to an annual check of appropriate parameters).

Baseline monitoring can involve sampling of reservoir bodies, streams or bores for data on what is considered normal for that water body (e.g., in metropolitan reservoirs of Western Australia the iron and manganese levels are monitored as part of their baseline programs, because of the presence of bauxite rich lateritic soils). In addition to monitoring for some key parameters (e.g., turbidity and pathogens) the monitoring of additional parameters such as hydrocarbons, pesticides and metals will depend on the activities occurring in the watershed; with monitoring occurring downstream of specific land uses and at the raw water point to determine what are baseline conditions. What to monitor for is closely linked to the land uses occurring in a watershed. Table 7.1 provides a guide to land uses and the resulting hazard sources, events and hazards.

In preparing a routine monitoring program, it's important to keep in mind that the greatest pollutant loads normally occur during storms. Baseline monitoring can miss major contamination incidents. Monitoring programs should be capable of capturing these events. Additionally, it is not always possible to monitor some parameters owing to financial constraints (e.g., one cannot monitor for all known pathogens). In addition, the logisitics and nature of the paramenter may make it difficult to monitor, e.g., cattle generally do not excrete *Cryptosporidium* at all times, only when they are infected; hence it is better to monitor for indicator organisms (such as faecal coliforms) and assume *Cryptosporidium* and other pathogens are present if the indicator is detected.

The importance of event-based monitoring programs, targeting events with the potential to impact water quality, cannot be understated, given most pollutants are washed off during heavy rainfall events (CRCWQ&T 2004). Any sudden or extreme change in water quality, flow or environmental conditions (extreme precipitation, flow rate, flooding) should arouse suspicion of adverse conditions that might cause drinking water

Table 7.1 Land uses and the resulting hazard source/event/hazard

Land Use/Activity	Hazard Source/Event	Hazards
Airstrip	• Storage/spillage of fuel/chemicals , mechanical servicing • Control of weeds/pests	• Hydrocarbons • Pesticides
Aquaculture Facilities	• Disposal of wastewater from ponds	• Nutrients
Camping, bushwalk-ing, cycling, rogaining, orienteering	• Human presence and possible bodily contact with water body • Damage to tracks or activities occurring off tracks	• Pathogens • Turbidity
Caravan Parks, campgrounds	• Wastewater system – septic tanks • Gardening maintenance	• Nutrients, Pathogens • Nutrients, pesticides
Cemetery	• Decomposition	• Nutrients
Defence Force activities	• Use of vehicles • Human presence, lack of toilet facilities	• Hydrocarbons, Turbidity • Pathogens
Depots— other utilities, shire or county	• Storage/use of hydrocarbons and pesticides and other chemicals	• Hydrocarbons • Pesticides
Dry Cleaners	• Storage and disposal of dry cleaning solvents	• Hydrocarbons
Farming/ Orchards / Wineries / Pastoral/ Market gardening	• Pesticide application • Fertiliser application • Manure application • Machinery/vehicles/ servicing/fuel storage • Livestock – pastoral animals • Housing – Septics • Land ploughing • Land clearing	• Nutrients • Pathogens, including *Cryptosporidium* and *Giardia* • Hydrocarbons • Turbidity, Pathogens including *Cryptosporidium* and *Giardia* • Pathogens, Nutrients, • Turbidity • Salinity
Feral animals	• Fecal contamination/animal carcasses • Wallowing etc • Human presence – hunting/trapping • 1080 baits are naturally occurring poison and create minimal risk	• Pathogens • Turbidity • Pathogens
Fire management	• Fuel reduction burns • Construction maintenance and usage of firebreaks and water points	• Turbidity • Turbidity, Hydrocarbons, • Pathogens
Fishing	• Human presence, often with dogs, in direct bodily contact with water and lack of toilet facilities • Vehicle presence close to water body	• Pathogens • Turbidity, Hydrocarbons
4WDs, motor cycles, unlicensed vehicles	• Use of unsealed roads and tracks, driving off roads/tracks • Fuel spills from refuelling or from accidents • Animal and human presence in the watershed and possible bodily contact with water body, lack of toilet facilities	• Turbidity • Hydrocarbons • Pathogens
Golf courses/ Grass tennis courts	• Septic tanks • Pesticide application • Fertiliser application • Gardens/manure application	• Pathogens • Pesticides • Nutrients • Pathogens

Land Use/Activity	Hazard Source/Event	Hazards
Horse riding/Animal exercising	• Animal and human presence in the catchment and possible bodily contact with water body, lack of toilet facilities • Uses of horses on sealed tracks and riding away from trails	• Pathogens • Turbidity
Housing	• Septic tanks • Gardens/manure application	• Pathogens • Nutrients, Pesticides, Pathogens
Illegal drug crops	• Human and animals (e.g., dogs) in the catchment and possible bodily contact with water body • Use of animal manure as fertilizer for crop • Control of weeds/pests	• Pathogens • Pathogens, Nutrients • Pesticides
Mechanical servicing/ Fuel stations	• Fuel spills / fuel storage • Septic systems	• Hydrocarbons • Pathogens, Nutrients
Mining	• Clear felling, ripping, ineffective site management, road construction/ upgrades • Machinery/vehicles/servicing/fuel storage • Offices — septic systems, human presence • Rehabilitation of land • Chemicals — will depend on type of mine (e.g., cyanide for gold) and processes occurring on the mine site	• Turbidity • Hydrocarbons • Pathogens, Nutrients • Nutrients, Pesticides • Toxic Chemicals
Motor rally events	• Use of unsealed roads, import of material to repair tracks • Fuel spills from servicing/accidents • Human presence (spectators, competitors and officials)	• Turbidity • Hydrocarbons • Pathogens
Non-powered boating (e.g., canoes, electric craft)	• Possible bodily contact with water • May be the need for a fuel powered rescue craft	• Pathogens • Hydrocarbons
Picnic areas, Reservoir buildings, Rangers homes	• Humans and animal (e.g., dogs) in the watershed and possible bodily contact with water body • Septic tanks in picnic facilities/Ranger homes • Chemical storage in Operations building	• Pathogens • Pathogens, Nutrients • Will depend on chemicals stored
Power stations	• Possible storage of large quantities of fuel and transportation of fuel, mechanical servicing	• Hydrocarbons
Production bores	• Fuel storage • Control of weeds/pests	• Hydrocarbons • Pesticides
Residential/Semi-rural	• Septic tanks • Vehicles, machinery • Gardens	• Pathogens • Hydrocarbons • Pesticides, Nutrients
Resource harvesting in native forest (e.g., apiaries, wildflower picking, seed collection)	• Human presence • Use of unsealed roads	• Pathogens • Turbidity

Land Use/Activity	Hazard Source/Event	Hazards
Rifle range	• Wastewater disposal • Disposal of bullets	• Pathogens • Lead
Roads and tracks/Gravel pits	• Erosion of unsealed roads and tracks • Fuel and chemical spills • Increased public access	• Turbidity • Hydrocarbons • Pathogens
Rubbish/Garbage pits/ Landfills	• Leaking of contaminants into groundwater	• Nutrients, depends on chemicals/ materials stored at site, e.g., batteries, cadmium, lead, pesticide containers, hydrocarbons
Shire/County roads, Infrastructure maintenance (power lines, telephone lines, towers, pipelines)	• Construction/Maintenance of roads, erosion of unsealed roads • Fuel and chemical spills • Weed control • The presence of motorists or their vehicles' loads	• Turbidity • Hydrocarbons • Pesticides • Pathogens
Siviculture (tree plantations)	• Application of fertilizers • Control of weeds/pests • Storage of chemicals/fuels • Also see risks associated with timber harvesting below	• Nutrients • Pesticides • Hydrocarbons
Sporting facilities (golf courses, ovals, tennis courts, race course, show grounds)	• Use of fertilizers • Use of herbicides/pesticides • Change rooms/toilets, Animals • Use of machinery, oiled sand greens at golf course	• Nutrients • Pesticides • Pathogens • Hydrocarbons
Swimming	• Humans, dogs, horses, etc. swimming in water body or feeder streams, lack of toilet facilities	• Pathogens
Timber harvesting	• Road construction/upgrading, construction of landings • Vehicles and machinery • Fire risk from machinery/human presence • Human presence and increased public access due to new roads/tracks • Spread of forest disease from vehicles • Chemicals/pesticides from harvesting activities	• Turbidity • Hydrocarbons • Turbidity • Pathogens • Turbidity • Pathogens • Pesticides
Wastewater treatment	• Disposal of effluent in catchment (e.g., water reuse)	• Pathogens, Nutrients

contamination (Hrudey 2003). Even where raw water storage is provided, sudden heavy precipitation can initiate changes in raw water quality that may impair the ability of the treatment processes to cope and achieve optimal treatment (Hrudey 2003).

The following is a list of the types of event that can trigger event based monitoring:

- Large river flow events (e.g., after a cyclone/hurricane)
- Significant rainfall event (e.g. >50mm [>2 inches] in 48 hrs)
- A natural disaster (e.g., wildfire, landslide)
- A drought event when water levels are low and dilution and detention are lower than normal
- A pollution incident (e.g., chemical/ hydrocarbon spill)
- An environmental event (e.g., fish kill in water body or feeder streams).

When conducting event-based monitoring, the measured concentrations of a parameter need to relate to flow as the total load is of interest (i.e., lb or kg of pollutant). Any monitoring program must have a clear objective — what is to be monitored, why, when and how will the data be used?

Ideally, results with high statistical power are needed (i.e., results with a high degree of certainty); power is related to sample size and the size of the effect being sought. Table 7.2 below provides a general summary of the interactions between these factors. However, the benefits of implementing a rigorous monitoring program are often outweighed by the lack of funds available for its implementation.

Frequently the resources are not available to implement a rigorous monitoring program and examples from the literature have to be found, assuming that similar responses will apply to the local situation.

Some US examples of success stories can be seen in Spooner et al. (1999). For an evaluation of the US Rural Clean Water Program, refer to the project examples at www.water.ncsu.edu/watershedss/info/rcwp/ (this site also has some valuable lessons on implementation).

These examples do show that it is possible to address some water quality issues in the short term and see actual improvements.

The timeframe to see measurable effects of management actions probably ranges into decades. Many of these problems have taken decades to become apparent and it seems likely that it will take many decades to overcome their effects. However, in some local cases, it may be possible to see quick results from a local management intervention. For example, in a small watershed, improvements in suspended sediment and turbidity could show quickly after fencing off the stream from stock (Spooner et al. 1999). Vegetative treatments, especially those involving trees, could take some years to affect shading, and thus stream temperature, or to slow down erosion.

Table 7.2 Sample size and statistical power

Effect Size	Sample Size	Power
Small	Small	Very Low
Large	Small	Low
Small	Large	Low
Large	Large	Very High

Monitoring Frequency

The frequency of water quality monitoring is linked to on-ground survey, the level of risk associated with land use activities in an area and historical water quality results. Where land uses in a drinking water watershed pose a high or severe risk to water quality, monitoring programs must be put in place that will stand up to expert scrutiny, e.g., cross examination in a coroner's court. A model to illustrate how sampling frequency increases as risk increases is shown in Table 7.3. The levels of risk that have been used in ascending order are are: trivial, low, moderate, significant, major, high and severe. These levels of risk will vary depending on the risk model used.

For sampling in surface water sources/watersheds, ideally sample the first significant flush of the wet season that reaches the reservoir body, the timing of additional sampling is dependent on the length of time the stream is flowing through the year.

For high risk areas frequent sampling on its own may not be sufficient. Other resources that may be required include: flow gauging stations, SCADA, and increased surveillance.

BEST PRACTICE

A monitoring program should include both microbiological and physical contaminants. The programs should have clearly defined objectives, generally aim at linking cause and effect and be robustly designed. The question of load based or concentration measurement needs to be

Direction of increasing risk

→

Direction of increasing frequency and scale of sampling

Table 7.3 Sampling frequency for parameters in watersheds

Risk level	Trivial	Low/ Moderate	Significant/ Major	High/Severe
Surface Water Sources	N/A	Annual	Quarterly	Monthly
Groundwater Sources	N/A	Annual	Quarterly	Monthly

considered. Other key points related to monitoring programs include:

- Calibration of sites, instruments and laboratories
- Quality assurance and quality control programs are essential
- Interpretation of results is critical (otherwise why collect the data?)
- Data integrity is a sensitive area
- Control is required over the dissemination of data.

Studies over the years have demonstrated that, for surface waters, impacts are usually directly related to rainfall-induced runoff, and therefore sampling during relatively dry periods provides minimal indication of the level of contaminants (CRCWQ&T 2004). For example:

- Load measurements showed as much as 300 years of dry weather pathogen contamination loads could be exported during one day in a single, *small* event
- During baseline, or small storm events, source water quality can be heavily degraded in agricultural or urbanised watersheds, highlighting the need to direct data collection more to high runoff periods than dry weather periods.

Proposed microbial qualitative risk assessment categories that can be used to determine monitoring frequencies for watersheds are detailed in Table 7.4.

Early warning water quality monitoring systems have been identified as valuable components of watershed management. In terms of groundwater this is particularly important because when a pollutant is detected in a bore it is too late; early warning systems are required that will fit groundwater systems. The USEPA is developing a system to link points to a database and input to a GIS-based early warning system in the future. Additionally, Melbourne Water and Ohio River Valley Water Sanitation Commission (ORSANCO) already have early warning systems in place.

CASE STUDY

Hypothetical surface water reservoir with a 5 gigalitre capacity — numerous land uses occur in the watershed (refer to map on Figure 7.1). Apart from farming properties the remainder of the watershed is state forest, the watershed is located 20 kilometres from a large town (population 100,000), recreation does not occur in the watershed.

Watershed Sampling

- At sample points 1, 2, 3, 4 and 5 sample for major components at least once a year. Given the intensive land uses occurring in this watershed; should sample first significant flush that reaches the reservoir body and two or three heavy rainfall events (e.g., >50 mm in 48 hr per wet season).
- Microbiological sampling to occur at sample points 1, 2 and 3 annually (due to moderate microbiological risk from septic tanks).
- Microbiological sampling to occur at sample point six-monthly (due to high

Table 7.4 Qualitative microbial risk assessment for watersheds

Impact Level	Source Class	Runoff Conditions		
		Dry Weather Baseline Conditions	Baseline Event — Small Event	Large Event — Extreme Event
Low (protected watershed)	Large Reservoir	Very Low	Low	Moderate
	Small Reservoir	Low	Moderate	Moderate – High
	River/ Stream	Low	Moderate – High	High
Moderate (partly impacted watershed)	Large Reservoir	Low	Moderate – High	High
	Small Reservoir	Low	High	High – Very High
	River/ Stream	Moderate	High	Very High
High (heavily impacted watershed)	Large Reservoir	Low – Moderate	High	Very High
	Small Reservoir	Moderate	Very High	Extreme
	River/Stream	High	Very High – Extreme	Extreme

microbiological risk from intensive feedlot cattle property).

- Sample pesticides and hydrocarbons at sample points 1, 2, 3 and 4, four times per year (due to widespread use of pesticides (major risk) and hydrocarbon risk [major risk] from farming properties).
- At sample points 1, 2, 3 and 4 sample for nutrients annually due to risk from cattle and fertiliser use.
- Turbidity probes placed below farming properties to monitor turbidity from farming activities.

WHAT SHOULD WE KNOW?

- How do you rationally determine appropriate frequencies for monitoring programs?
- Are there any computer-based packages for source protection operations data (from surveillance) with linkage to drinking water quality programs?
- Value of raw water pathogenic monitoring versus recognised risk (e.g., Water Corporation is the only major water utility in Australia not currently testing for *Cryptosporidium* and *Giardia*).

REFERENCES

Cooperative Research Centre for Water Quality and Treatment. 2004. 'Pathogen movement and survival in catchments, groundwaters and raw water storages', *Fact Sheet 8: Catchment tributary microbiology: rain events versus dry baseline periods.*

Figure 7.1 Hypothetical watershed

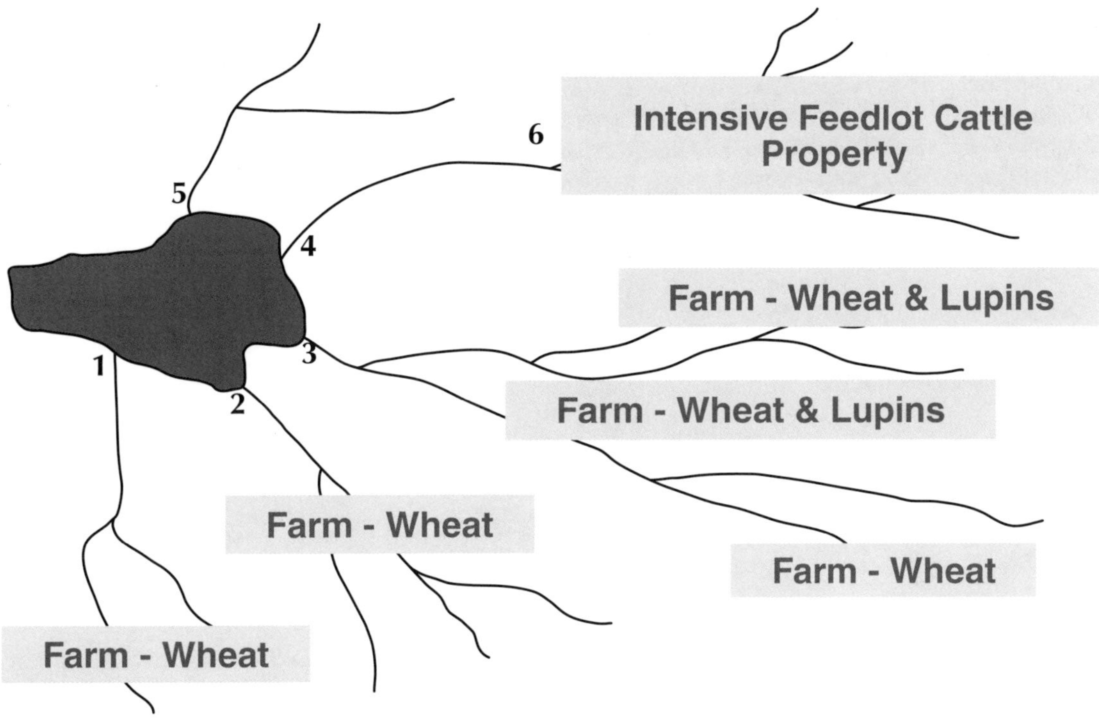

Risk/Vulnerability Assessment Results
Farms — Wheat/Lupins
- Pesticides: Major risk
- Hydrocarbons: Major risk
- Septics: Microbiological, Moderate risk
- Farming practices: Turbidity, Major risk
- Fertiliser use: Nutrients, Major risk

Farms — Cattle
- Cattle: Microbiological, High risk
- Cattle: Nutrients, Major risk
- Hydrocarbons: Major risk
- Septic systems: Microbiological, Moderate risk
- Farming practices: Turbidity, Major risk
- Sampling Program required
- Raw water offtake from reservoir
- Microbiological: Weekly
- *Basic components: Monthly
- **Major components: Quarterly with DO and Temp profile
- Pesticides: Quarterly
- Hydrocarbons: Quarterly
- Nutrients: Quarterly

* Basic component suite = Aluminium unfiltered, True Colour, Conductivity at 25C, Iron unfiltered, Manganese unfiltered, pH-laboratory measure and turbidity

** Major component suite = Basic components (above) + Alkalinity as $CaCO_3$, Alkalinity as HCO_3, Calcium, Chloride, Filterable Organic Carbon, Hardness as $CaCO_3$, Magnesium, Nitrite plus Nitrite as N, Potassium, Silicon as SiO_2, Sodium, Sulphate, total filterable solids by summation.

CRCWQ&T See Cooperative Research Centre for Water Quality and Treatment.

Hrudey SE. 2003. 'Key Messages for Drinking Water Providers from Walkerton, North Battleford and Other Major Outbreaks,' *Conference Proceedings AWA Ozwater Conference,* Perth, Western Australia.

Peel MC, McMahon TA. 2001. *Managing Nutrients in Irrigation Drains — are we making a difference — will we be able to tell if we are making a difference?* Stage 1. Final, Centre for Environmental Applied Hydrology (CEAH) The University of Melbourne, Report for Goulburn-Murray Water.

Spooner J, Line D, Grabow G, Lombard L, Tweedy K, Humenik F, and Osmond D. 1999. 'Contributions and successes of long-term watershed programs in the United States: the rural clean water protecion program and the Section 319 national monitoring program.' *Proceedings of the International Conference on Diffuse Pollution*. Perth, Western Australia, May 1999. International Association on Water Quality.

8 RECREATIONAL ACCESS TO SOURCE WATER RESERVOIRS & THEIR WATERSHEDS

Michael A. Anderson, Rachael Miller, Perri Standish-Lee

CONTEXT

Increasing demand for recreational opportunities, especially during the hot summer months, and a depletion of wilderness areas close to major population centres have put public and political pressure on water purveyors to make every body of water available for multi-purpose use. Particular interest exists for direct use of the surface water supply for boating, fishing, water skiing and swimming. Recreational use of the watershed for hiking, camping, off-road vehicle use, horseback riding and biking is also sought in many instances.

Such recreational activities can have direct and indirect impacts on water quality, however. These can reduce the quality, safety and supply security of the source water, which, in turn, puts a higher expectation on downstream water quality barriers such as water treatment. Other issues besides public health and safety can also be impacted by recreational access, including ecology, sustainability and security.

Water-based Recreation

Recreational activity has been conclusively shown to have negative impacts on drinking water quality (CRCWQT 2006). For example, boating has been shown to introduce MTBE, benzene, toluene and other contaminants into the water column (USEPA 1996; Boughton and Lico 1998; Baehr and Zapecza 1998). Two-stroke outboard motors in particular can release significant amounts of fuel into the water column; since the intake and exhaust valves are open concurrently when the air-fuel mixture is drawn into the combustion chamber, up to 25 percent to 30 percent of the unburned fuel passes through the engine and is exhausted directly into the water, where the more soluble components can then dissolve into the water column (Junter et al. 1995). Methyl-*tert*-butyl ether (MTBE), a potential carcinogen that also imparts an undesirable taste and odor to water, has been found in a number of lakes and reservoirs as a result of recreational watercraft use (An et al. 2002; Dale et al. 2002; Boughton and Lico 1998). Benzene, toluene and xylenes (known carcinogens) have also been found in surface waters subject to motorized recreational watercraft use. Inputs from motor operation, as well as spills during fueling and leaks on boats and fuel storage facilities, all contribute fuel to surface waters (USEPA 1996; An et al. 2002).

Recreational activities that involve direct body contact with the water, such as swimming, are potentially a greater cause of concern for many water utilities. While studies demonstrating significant degradation of water quality in source water reservoirs as a direct result of body contact recreation are limited, substantial microbial contamination has been found in many water parks and other recreational waters that annually result in a large number of outbreaks of gastroenteritis and non-gastroenteritic diseases within the recreational community. For example, a total of 65 waterborne disease outbreaks associated with recreational water were reported in 23 states within the United States in 2001–2002 (Yoder et al. 2004). These outbreaks caused illness in 2,536 individuals, resulting in 61 patients being hospitalized and eight deaths (Yoder et al. 2004). Poor surveillance and reporting practices suggest that actual numbers are probably much higher, however. Of these 65 outbreaks, 30 involved gastroenteritis, with the etiological agent identified in 23 of them. *Cryptosporidium spp*. accounted for 16.7 percent of the 12 outbreaks in fresh water (e.g., lakes) and 50 percent of the 18 reported outbreaks in treated waters, such as swimming pools (Yoder et al. 2004). *E. coli* O157:H7, norovirus, *Shigella sonneii* and *Giardia intestinalis* were also identified as etiological agents in the outbreaks (Yoder et al. 2004).

Cryptosporidiosis surveillance in the United States for the period 1999–2002 revealed a strong seasonality, with a six-fold increase in the number of cases reported in July and August relative to the number of cases reported during the winter months (Hlavsa et al. 2005). A marked increase in cases of giardiasis was also found during the summer months (Hlavsa et al. 2005b). These increases were more dramatic for young children aged 1–4 years and 5–9 years and that is thought to reflect increased use of communal swimming venues (e.g., lakes, rivers and swimming pools) during the summer recreational season (Hlavsa et al. 2005a, 2005b). The large percentage of outbreaks due to *Cryptosporidium* in treated waters (50 percent) underscores the resistance of this organism to conventional treatment.

These observations thus have direct implications for water utilities; that is, body-contact recreational use can result in microbial contamination of source water supplies with organisms that resist conventional treatment. These findings also indicate that body-contact recreational use can result in increased illness, especially amongst the recreational community, by a wide range of pathogenic organisms, including *E. coli* O157:H7 and *Shigella sonnei* (Yoder et al. 2004).

Land-based Recreation

In addition to the impacts of recreational activities directly on the surface source water supply, activities within the watershed can also potentially degrade water quality. Animal waste, such as from dogs and horses, even away from the lake, reservoir or river, can be flushed into the receiving water body during periods of rainfall and runoff. Moreover, sanitary facilities can leak and release microbial contaminants to the surface and subsurface that may find their way into the water supply. Even when sanitary facilities are fully functional and pet waste is eliminated from the watershed, hiking and biking can result in accelerated erosion of soil within the watershed, and that can introduce nutrients, pathogens and sediment to the source water supply. Sediment in particular can also reduce the effectiveness of downstream chlorination. The complex interrelationships that characterize watershed processes make it difficult to provide a quantitative estimate of the extent of impact, given an anticipated or measured quantity of recreational access (CRCWQT 2006).

Permitting recreational access is potentially at odds with efforts at source water protection; the weakening of this first barrier (by reducing the watershed's ability to mitigate risk, as well as contributing more risk) may necessitate higher levels of treatment to compensate for potentially poorer source water quality. Treatment barriers, which are known to fail, only reduce contaminant concentrations, so residual risk persists (CRCWQT 2006).

Risk Assessment

To investigate the potential impacts of recreation at reservoirs used for a potable water supply, a number of different types of control strategies can be considered. Before any strategy is adopted, however, a careful assessment is needed of the hazards, barriers and water quality of the water supply system, from watershed to tap/faucet. Ideally, a risk assessment mechanism should be used which outlines hazards, hazardous events, barriers, quantitative information on hazards and barrier performance and a measure of certainty (of information). This will then assist in management decisions by identifying key risks and areas for mitigation action. The use of the risk-based approach also addresses due diligence.

CRCWQT (2006) suggests that key issues that influence the categorisation of risk can include:

Likelihood Criteria

- Level of contact with the water source
- Location of the activity in the storage or watershed
- Frequency of use and number of recreators
- Time (duration) spent in the watershed
- Infectivity of recreators
- Behaviour of recreators
- Contaminant release time
- Availability and use of watershed-based facilities (i.e., toilets and bins)
- Existing management controls and their effectiveness
- Characteristics of the water source (vulnerability)
- Watershed land condition (vulnerability)
- Impact on long-term water quality condition
- Impact on long-term watershed condition.

Consequence Criteria

- Direct human health impacts
- Indirect human health impacts
- Level of recreational use
- Potential to cause a breach of water quality standards
- Impact on water supply operations (interruptions)
- Criticality of the source (risks that affect direct source supply)
- Management options and their effectiveness.

Useful Tools

To support the risk assessment process in determining the appropriate level of management control for an activity or total watershed, several tools can be used to move from a more qualitative approach to a semi or fully quantitative approach. These include:

- Water quality sampling programs — compliance, investigations, seasonality, etc.
- Hydrographic sampling programs
- Reservoir dynamics
- Spatial information – land uses, slope, boundary information, rainfall zones, point and diffuse contaminant sources
- Models
- Nonspatial information — research outcomes, recreation activity information, audit information, etc.
- Barrier performance information, e.g., short-circuiting and water treatment failures.

With this information, a comprehensive understanding can be deduced of the seasonal and spatial impacts on water quality and public health that would likely result from a range of recreational activities. Depending upon the situation, a smaller subset of properties and factors can be sufficient to understand potential water quality and public health effects.

OPTIONS

Options to limit potential negative impacts from recreation include:

Prevention or elimination of recreation from the source water supply and watershed—In some cases, it may be necessary to eliminate recreation, especially body-contact recreation, from the source water supply. While this is seemingly the most straightforward approach to minimizing potential negative impacts on water quality from recreational activity, it has key social implications. As such, a thorough analysis of the impacts, and a careful assessment of alternative actions, including increased treatment, would generally be necessary to defend such an action. Ranger patrols are usually required to enforce these closures. However, if alternative areas can be provided for the activity, then these can be pursued.

This can include separate swim lagoons. Such a facility thus maintains the recreational activity near the reservoir, but removes this body-contact recreation from the source water supply. The recreational experience for beach-goers is quite similar to that found at the supply, but with some significant benefits. First of all, since such swim lagoons typically have filtration and chlorination, the risk of infection to the recreators is potentially much lower than the untreated beach area. Moreover, the improved water clarity that results from filtration and disinfection provides a safer swim environment; for example, improved water clarity allows lifeguards and others to keep track of those swimming in the lagoon, and respond to anyone that may be struggling. Related to that, the confines of the swim lagoon necessarily keep anyone from swimming out too far from shore and into deep water. Nevertheless, this will only reduce the total number of people participating in body-contact recreation when water-skiing and personal watercraft (PWC) use still occur on the main reservoir.

Restricted access to areas and seasons that are less vulnerable to contamination—In some watersheds, areas can be identified that have low connectivity to the off-take of the raw water source and, as such, are less vulnerable to the impacts of recreation. Also, depending on climate, some seasons are drier and have less likelihood of runoff, allowing opportunity for land-based recreation. Other reservoir use designs can also reduce the impact of direct body-contact recreation, boating, and other activities. For example, it may be possible to limit contamination at the water intake by restricting recreational use to regions of the reservoir, lake or river that are far removed from the intake.

This may allow sufficient dispersion and loss to lower the microbial and/or chemical contaminant levels delivered to the treatment plant. The effectiveness of this as a mitigation measure requires detailed information about the hydrodynamics, transport and fate of contaminants within the system that may only be available through numerical modeling. Stratification during the summer may help isolate contaminant inputs from a hypolimnetic intake, since effectively all recreation occurs on or very near the lake surface, although the pathogens and other contaminants associated with particulate matter may settle through the thermocline and thus potentially still reach the intake.

Permit-based access—Permit-based access can limit the number of users to the watershed. This can be particularly useful for swimming and boating activities or club-based hiking, horseback riding, and related activities. Permits are issued with specific drinking water quality-based conditions such as "waste in, waste out" which are audited randomly by rangers. A breach of conditions can result in a loss of permit.

Maintenance of active water quality surveillance and closure policies—Regular water quality monitoring will help quantify the water quality in the source water supply and can identify chronic contamination events. Closure of the facility may be justified under conditions of high contaminant concentrations. Such a strategy is of little use for acute contamination events, however, since a substantial lag often exists between sampling and when analytical results are available. This is especially true for standard microbial assays that often require 72 hours or more before results are available. This approach may become more effective as near real-time techniques are developed.

Establish a maximum allowable number of recreators that is protective of water quality—It may be possible in a general way to define a maximum number of boaters, swimmers, water-skiers and others that can participate in recreation on the source water supply without compromising finished water quality or water consumers' public health. To do this requires understanding all sources of contaminants to the water supply, the concentrations at the intake and that reaching the treatment plant, and the removal efficiency at the plant. Modeling will most likely be needed to identify the maximum allowable recreation activities and their location relative to the intake.

Construction of a new intake that reduces contaminant incursions—In cases where recreational use is not readily moved, it may be possible to construct a second raw water intake located upstream and/or separated from the area of the water body used for recreation. The second intake, for example, could be used until reservoir levels drop to a level that requires use of the downstream main intake located at the dam. This would allow recreation to occur during the heavy use summer season; recreation could be prohibited in the fall (autumn) and winter months when the main intake is used.

Regular inspection and maintenance—Provision of regular inspection and maintenance of all sanitary facilities, fueling stations and other recreational service facilities should minimize the frequency and duration of leaks and other sources of contamination.

Public education—Educating the recreators, local community and drinking water customers of the importance of safe drinking water and the resultant management of the watershed and how that relates to recreational access is a key tool for successful recreation management.

BEST PRACTICE

Best practice in the management of recreation in water supply reservoirs is to err on the side of caution. Recreational use without a rigorous risk assessment is not recommended. A quantitative assessment is preferable to a purely qualitative approach; and, if the risk assessment demonstrates an unacceptable level of risk (e.g., more than one infection per 10,000 consumers per year) then either recreation must be prohibited or one of a range of management options must be implemented to reduce the risk to an acceptable level.

For many systems and activities, the ability to undertake a full quantitative risk assessment can be limited by available information. The complex interrelationships that characterise watershed processes make it difficult to provide a quantitative estimate of the extent of impact given an anticipated or measured quantity of recreational access (Miller et al. 2006). Moreover, uncertainties are always attached to predictive tools or models. The result of this irreducible uncertainty is that decisions must be made based on principles and probabilities (risk assessments).

The suite of management interventions that can be applied includes a range of options, all explained in more detail elsewhere in this chapter:

1. Prohibition of land- and/or water-based recreation, or water-contact recreation
2. Access restrictions based on location in the water body, or on the season
3. Permit-based access
4. Surveillance and monitoring; leading to closure in response to any deviations
5. Setting maximum numbers of recreators
6. Provision of dedicated recreation facilities on the site, e.g., a separate swimming pool (this is expensive but can provide win–win outcomes)
7. Relocating intakes or other structures
8. Regular inspection and maintenance
9. Public education about the links between recreation and community health.

In view of strong public — and hence political — pressure for recreational access, water utilities must be willing to stand firm to ensure that water quality is not compromised. Equally, considerations and analyses must be transparent and accessible, to ensure accountability.

A strong partnership between planners, the water utility, land managers and the health system must be established and maintained to ensure maximum collaboration and, in particular, optimum communication on health outcomes and health risks.

CASE STUDIES

North America

Recreation on source water supplies has been an issue of some concern in California. Three important case studies there highlight the intricacies of the issue, including the scientific and public health aspects, as well as the social and political concerns.

Diamond Valley Lake, California

The first case study focuses on Diamond Valley Lake (formerly known as the Domenigoni Reservoir and also the Eastside Reservoir), near Hemet, California. The reservoir is owned and operated by the Metropolitan Water District of Southern California (MWD); planning for the reservoir began in the 1980s, when it was recognized that the raw water storage capacity for the region was inadequate, especially in the face of population growth, possible droughts and natural disasters (e.g., earthquakes). Construction of the 1,800 ha reservoir began in 1995 and continued for 4 years at a cost of over US$2B (MWD 2006). During initial stages of planning and construction, there was vocal public support for full recreational access and use of the reservoir, including activities such as boating, fishing, waterskiing, personal watercraft (PWC) and swimming. Although MWD did not permit body-contact recreation at any of its other reservoirs, it agreed to investigate the public health implications for water consumers (Stewart et al. 2002).

The health implications from consuming water from a recreationally impacted lake were not well established, however, so a numerical model was developed that accounted for pathogen inputs from body-contact recreation, transport through advection and dispersion, and loss due to inactivation and settling (Anderson et al. 1998). The model could thus be used to predict concentrations at the intake, and further loss during transport to the plant and loss through conventional treatment, such that concentrations at the tap could also be estimated. Those predicted concentrations were then used with dose-response models to predict water consumer risk levels as a function of different types, distribution and extent of recreational activities on the lake (Stewart et al. 2002). Monte Carlo techniques were incorporated into the model to account for uncertainties in model inputs that allowed results to be expressed in terms of probabilities (Anderson et al. 1998).

Because of its long environmental half-life under many conditions and its resistance to chlorine disinfection, *Cryptosporidium* emerged as the pathogen of greatest concern. The model demonstrated that restriction of recreation to the eastern portion of the lake, away from the intake, lowered *Cryptosporidium* concentrations in water delivered to the treatment plant by up to 60 percent. The model further allowed estimates of the total number of recreators that could be accommodated on the reservoir without exceeding the USEPA's target of one infection per 10,000 consumers per year, although such activities were recognized to increase the overall risk of infection relative to the baseline of no body-contact recreation (Stewart et al. 2002). Further economic analyses suggested that body-contact recreation would add US$7.5 milion to US$13.9 million, or 1.8–3.3 percent, to the economic activity for the region, while the additional treatment plant capital costs to offset the increase in pathogen concentrations would be US$150m–US$190m and annual operating costs would be US$3.8 million to US$5.3 million (Stewart et al. 2002). Based on these analyses, the Board of Directors of the MWD voted against allowing body-contact recreation on the reservoir.

There were separate concerns about fuel, especially MTBE, in the new reservoir if a significant amount of fishing and other motorboat activity was allowed. Nearby Lake Perris, California, had very high levels of MTBE in the epilimnion of the lake (epilimnetic concentrations exceeded 20 µg/L in the summer of 1996) that was attributable to the high amount of motorboat activity there (Yates et al. 1997; Dale et al. 2000).

A separate modeling analysis was conducted to predict potential MTBE concentrations in Diamond Valley Lake, based upon expected motorboat demand. For this analysis, a simple analytical model was developed and validated for Lake Perris. MTBE has a very low partition coefficient to sediment or suspended solids and is not subject to photolysis or other transformation reactions, so MTBE concentrations were a function of inputs from motorboat activity and losses through volatilization. The model was then applied to the new reservoir, where MTBE concentrations were predicted to be lower than the levels found in Lake Perris and range from 9 to 13 µg/L during the summer recreational season (Yates et al. 1997).

The use of MTBE-free fuel and the banning of conventional two-stroke motors substantially lowered the predicted MTBE levels. Since MTBE is not removed to any substantial extent during conventional treatment, MWD decided to ban conventional two-stroke motors on the reservoir and also to require MTBE-free fuel be used in all motor-powered watercraft. Moreover, only non-contact boating (e.g., fishing) is allowed on the lake (no personal water contact or water-skiing is allowed).

While the local community was disappointed with the outcome of these studies and with the MWD Board's decision to ban body-contact recreation, the reservoir nevertheless is an important recreational resource for the region. MWD has employed a fisheries biologist at the reservoir since early in the construction process; with careful planning, he and the MWD have been able to develop a world-class fishery in a very short period of time that is expected to soon yield state and world-record size largemouth bass and other sport fish. Additional recreational facilities are also planned for the area, including a 75-acre boating and swimming lake (MWD 2006).

Lake Tahoe, California and Nevada

Lake Tahoe is known worldwide for its crystal clear waters, although increased recreational use and residential and urban development within the watershed threaten its water quality (Lico 2004). The lake is also used as a source of drinking water supply. Volatile organic compounds (VOCs) were detected in water samples collected in 1997 and they included the BTEX compounds, as well as MTBE and tert-amyl methyl ether; and were directly related to the amount of boating there (Boughton and Lico 1998). Polyaromatic hydrocarbons (PAHs) were also found in water samples (Lico and Pennington 1999).

On the basis of these and other monitoring studies, the Tahoe Regional Planning Agency enacted an ordinance that banned all carbureted two-stroke motors from Lake Tahoe effective as of June 1, 1999. Monitoring conducted by the USGS between 1999 and 2001 found a 90 percent reduction in MTBE levels after the ban (Lico 2004). The concentrations of other VOCs exhibited 10 to 60 percent reductions within Lake Tahoe and other lakes in the basin following the ban, although PAH concentrations were unchanged (Lico 2004).

Contra Loma Reservoir, California

The Contra Loma Reservoir is a small (32 ha) emergency supply reservoir located near Antioch, California. The reservoir, constructed by the US Bureau of Reclamation in 1967, is owned by the Contra Costa Water District (CCWD) and located within the East Bay Regional Park District's (EBRPD) 314 ha Contra Loma Regional Park. As part of the Park, the lake was open to the public for swimming, boating and fishing from 1968, although concerns over the public health impacts to water consumers from body-contact recreation led the California Department of Health Services (DHS) to order CCWD to either close the reservoir to body-contact recreation or cease use of the reservoir as a source of water supply (American City and County 2003). That compliance order thus directly pitted the water utility (CCWD) against the recreational interests (i.e., EBRPD).

Extrapolation of results from the Diamond Valley Lake study (described above) using recreational and reservoir data for the Contra Loma Reservoir, combined with separate hydrodynamic modeling, confirmed that body-contact recreation on this small reservoir posed a substantial threat to both downstream water consumers of the treated water as well as the swimmers themselves (Anderson 1999a,b). Discussions ensued about a compromise that could meet the high regional demand for body-contact recreation, while at the same time protecting the reservoir from pathogen inputs from such activities and meeting the DHS compliance order.

The concept of a swim lagoon that physically isolated the swimmers from the reservoir but provided a recreational experience similar to that found on the beach there emerged as an acceptable compromise. Modeling further demonstrated that the swim lagoon afforded greatly improved conditions for the swimmers, lifeguards and others there. Filtration and chlorination of the swim lagoon was predicted to lower the overall risk of infection to swimmers by 286-fold through chlorine inactivation of viruses and removal of protozoan (oo)cysts through filtration (Anderson 1999b). Construction of the swim lagoon began in September 2001 and was completed in May 2002. Construction costs were borne by CCWD, while the swim lagoon is owned and operated by the EBRPD. Since the swim lagoon opened on May 25, 2002, attendance at the park has increased significantly. This can be attributed in part to the water treatment processes that also greatly improve the clarity of the water relative to that in the reservoir. This clearer water is aesthetically and microbiologically cleaner, and also safer. That is, with improved clarity and a confined swimming area, parents are more comfortable with their children swimming there, and lifeguards can more readily locate anyone in difficulty.

Since then, treated swimming lagoons have been proposed for other multiple-use reservoirs in California and elsewhere.

Australia

Recreational access in watersheds varies across Australia, from closed watersheds, through limited access and up to open access. In most cases, recreation access is not permitted on direct supply sources, but is allowed on pump-backs or emergency sources. Some form of land-based recreation is often permitted in areas some distance from the reservoir or off-take.

Adelaide, South Australia

Several reservoirs supply water to the City of Adelaide. All reservoirs supplying drinking water to Adelaide have very limited recreational access. Swimming, boating and fishing are prohibited at all reservoirs except for one, which has a small fishing trial. Some passive land-based recreation is allowed in areas some distance from the reservoir. However, camping and hunting are also prohibited. Water treatment is chlorination, filtration and UV disinfection.

Access is limited through surveillance, education and the provision of alternative areas to recreate. As Adelaide is a coastal city, the distance to travel to the coast for water-based recreation is short.

Sydney, New South Wales

Sydney is supplied predominantly from one large reservoir, Lake Burragorang, behind the Warragamba Dam, and then several smaller reservoirs to the south of the city that operate as reserve storages. As Warragamba is a very large reservoir with an equally large watershed, some restricted land-based access can occur in areas with low connectivity to the reservoir. Rangers patrol watersheds to ensure recreational activities comply with approval conditions, and to identify and remove illegal recreators. For the reserve storages to the south of Sydney, fishing and passive boating is permitted. All sources have water treatment involving direct filtration and chlorination.

REGULATION

North America

Section 319 (USC 33 1329) of the Clean Water Act requires that individual states develop programs and management measures to reduce human caused non-point sources of contamination. As a result, the regulations regarding recreational activities on source drinking water supplies vary significantly. For example, the California Health and Safety Code prohibits body-contact recreation in reservoirs where water is stored for domestic use, but makes exceptions for the reservoirs in San Diego county, the Modesto Reservoir in Stanislaus County, the Nacimiento Reservoir in San Luis Obispo County, and the reservoirs of the State Water Project (to the extent that such activity is compatible with public health and safety requirements). Regulations explicitly excluding direct body-contact with a source water reservoir in California can be traced to at least 1913, when the Raker Act (H.R. 7207) was passed by the California House of Representatives that specified that "No person shall bathe... or in any way pollute the water within the limits of the Hetch Hetchy Reservoir."

The American Water Works Association (AWWA) recently reaffirmed its policy statement regarding recreational use of domestic water supply reservoirs, noting that protection of public health and drinking water quality should be given the highest priority in operational decisions for multiple-use reservoirs (AWWA 2004). Body-contact recreation and use of polluting two-stroke gasoline (petrol) motors on boats should be discouraged (AWWA 2004).

Australia

In Australia, drinking water quality is guided by the *Australian Drinking Water Guidelines* (ADWG) and its risk-based Framework, which outlines the key aspects of drinking water quality management across all accountable stakeholders and outlines strategy, planning, operations and research. Water utilities that deliver treated water to customers report to health authorities on their performance against the ADWG.

The ADWG emphasises the need to avoid waterborne disease outbreaks through the "protection of catchments [watersheds] from human and animal waste as a priority."

WHAT SHOULD WE KNOW?

Knowledge in the area of recreational use of water supply reservoirs is improving all the time, but there is still scope for better collation of data relating to this topic. Improved data on infection rates would be helpful, especially as different diseases are reportable (or not) in different jurisdictions, allowing many cases to potentially escape the notice of health practitioners.

OTHER RESOURCES

Many organizations throughout the world are concerned with drinking water quality and public health. Some of those organizations are listed below, with their web site links:

- World Health Organization (WHO), www.who.int/en/
- International Water Association (IWA), www.iwahq.org
- International Water Resources Association (IWRA), www.iwra.siu.edu/
- United States Environmental Protection Agency (USEPA), www.epa.gov/
- American Water Works Association (AWWA), www.awwa.org/
- American Water Resources Association (AWRA), www.awra.org/

- Water Environment Research Foundation (WERF), www.werf.org
- Australian Water Association (AWA), www.awa.asn.au/
- Cooperative Research Centre for Water Quality and Treatment (CRCWQT), www.waterquality.crc.org.au (This organisation was replaced by Water Quality Research Australia in July 2008).

REFERENCES

American City and County. 2003. *Lagoon preserves multi-purpose res*ervoir, www.americancityand-county.com/mag/government_lagoon_preserves_multipurpose/index.html (accessed 22 July 2006).

American Water Works Association. 2004. 'Recreational Use of Water Supply Reservoirs,' in *Statements of Policy on Public Water Supply Matters*, www.awwa.org/about/oandc/officialdocs/AWWASTAT.cfm (accessed 22 July 2006).

An YJ, Kampbell DH and Sewell GW. 2002. 'Water quality at five marinas in Lake Texoma as related to methyl-tert-butyl ether (MTBE)', *Environ. Pollut.* Vol. 118, no. 3, pp. 331–336.

Anderson MA, Stewart M, Yates MV, Gerba CP and Rose JB. 1998. 'Modeling the impact of body-contact recreation on pathogen concentrations in a source drinking water supply', *Water Res.* vol. 32, no. 11, pp. 3293–3306.

Anderson MA. 1999a. *Estimated Pathogen Concentrations in the Contra Loma Reservoir and Associated Risk to Consumers and Recreators*, Final Report to the Contra Costa Water District.

Anderson MA. 1999b. *Hydrodynamic and Transport Simulations of the Contra Loma Reservoir: Predicted Pathogen Concentrations at the Beach and in the Proposed Swimming Lagoon*, Final Report submitted to the Contra Costa Water District.

Baehr AL and Zapecza, OS. 1998. 'Methyl *tert*-butyl ether (MTBE) and other volatile organic compounds in lakes in Byram Township, Sussex County, New Jersey, summer 1998', *U.S. Geological Survey Water-Resources Investigation Report* 98–4264.

Boughton CJ and Lico MS. 1998. 'Volatile Organic Compounds in Lake Tahoe, Nevada and California, Water Year 2000', *U.S. Geological Survey Water-Data Report NV-00-1.*

Cooperative Research Centre for Water Quality and Treatment. 2006. *Recreational Access to Drinking Water Catchments and Storages in Australia*, Research Report 24, Adelaide, South Australia.

CRCWQT — see Cooperative Research Centre for Water Quality and Treatment.

Dale MS, Koch B, Losee RF, Crofts EW and Davis MK. 2000. 'MTBE in southern California water — Motorized recreational watercraft were the predominant mode of contamination in reservoirs', *Jour. AWWA* vol. 92, no. 8, pp. 42–51.

Hlavsa MC, Watson JC & Beach MJ. 2005. 'Cryptosporidiosis surveillance — United States 1999-2002', in: *CDC Surveillance Summaries*, January 28, 2005. *MMWR* 54(SS01), pp. 1–8.

Hlavsa, MC, Watson, JC and Beach, MJ. 2005. 'Giardiasis surveillance — United States 1998-2002', in: *CDC Surveillance Summaries*, January 28, 2005. *MMWR* 54(SS01), pp. 9–16.

Jüttner FD, Backhaus U, Matthias U, Essers, Greiner R and Mahr B. 1995. 'Emissions of two- and four-stroke outboard engines — I. Quantification of gases and VOC', *Water Res.* Vol. 29, no. 8, pp. 1976–1982.

Lico MS. 2004. 'Gasoline-related organics in Lake Tahoe before and after prohibition of carbureted two-stroke engines', *Lake Reserv. Manage.* vol. 20, pp. 164–174.

Lico MS and Pennington N. 1999. 'Concentrations and Distribution of Manmade Organic Compounds in the Lake Tahoe Basin, Nevada and California. 1997–1999', *U.S. Geological Survey Water-Resources Investigations Report 99-4218.*

Metropolitan Water District of Southern California. 2006. *Diamond Valley Lake – the Jewel of California Lakes*, http://www.dvlake.com (accessed 22 July 2006).

Miller R, Bennett B, Birrell J and Deere D. 2006. *Recreational Access to Drinking Water Catchments and Storages in Australia,* Research Report No. 24, CRC for Water Quality and Treatment, Australia.

MWD — See Metropolitan Water District of Southern California.

Stewart MH, Yates MV, Anderson MA, Gerba CP, Rose JB, DeLeon R and Wolfe RL. 2002. 'Predicted public health consequences of body-contact recreation on a potable water reservoir', *Jour. AWWA* vol. 94, no. 5, pp 84–97.

US Environmental Protection Agency 1996, *Managing Nonpoint Source Pollution from Boating and Marinas*, EPA-841-F-96-0041.

Yates MV, Anderson MA, Gerba CP and Rose JB. 1997. *The Impact of Body-Contact Recreation in the Eastside Reservoir on 1) Pathogen Risk to Consumers and Recreators and 2) MTBE Contamination,* Final Report to the Metropolitan Water District of Southern California.

Yoder JS, Blackburn BG, Craun GF, Hill V, Levy DA, Chen N, Lee SH, Calderon RL and Beach MJ. 2004. 'Surveillance for waterborne-disease outbreaks associated with recreational water – United States, 2001-2002', In: *CDC Surveillance Summaries*, October 22, 2004, *MMWR* 53(SS08), pp. 1–22.

9 ON-SITE WASTEWATER TREATMENT SYSTEMS

Robert (Bob) Ford

CONTEXT

On-site wastewater treatment systems accept wastewater from residential and commercial properties and small developments. They treat domestic wastewater and dispose of the residual liquids within the property concerned. A characteristic of these systems is that the responsibility for their operation and maintenance lies with the individual property owner or an owners' committee, rather than a public entity or water company. The owner may outsource these tasks to an external contractor, but the legal responsibility for the unit's performance rests with the owner.

Such systems are usually small domestic-size treatment plants or septic tanks with a disposal area. The risk to water quality in the watershed comes from the cumulative impact of a large number of these small systems, their density and varied performance. There will usually be a wide variability in their performance, which will depend on the appropriateness of the system for each site, its age, level of maintenance, current property owner's understanding and the owner's willingness and ability to comply with the relevant regulations and the manufacturer's recommendations. Generally, owners are unwilling to spend money on maintenance unless compelled to do so by a statutory authority, or a failure occurs that results in significant inconvenience to them or their family.

Consequently, it can be expected that a watershed that includes unsewered properties will have a population of small treatment systems, a percentage of which are not performing to specification and may be discharging polluted and infectious material into the environment.

As these systems all discharge nutrients and usually pathogens it is likely that, even if a high percentage of units are operating satisfactorily, the local density of the development may be sufficient to pollute the local groundwater.

OPTIONS

There are a number of issues that need to be considered in establishing "Best Management Practice" for on-site systems in a watershed. They include:

1. A process for determining the suitability of any site for on-site disposal of wastewater. This includes a policy and willingness to refuse approval for a site that is too small or considered unsuitable for development
2. An appropriate process for the design and selection of the system (including the treatment and effluent discharge/disposal method) for each site, having regard to the site and its constraints and

ability of the current and future owners to operate and maintain the selected system
3. A compliance regime that aims to educate, encourage and, if necessary, compel current and future owners to operate and maintain their systems in accordance with the manufacturers' requirements and the approval conditions
4. A program that monitors and records performance, detects failures and enforces rectification
5. A process that establishes local density limits for these systems and prevents an excessive concentration likely to pollute groundwater and surface waterways
6. A licensing and recording system for assessing applications for new systems, setting operating conditions (where approval is granted), enforcing compliance and recording ownership and system details.

These actions may be carried out by a variety of agencies.

Risk Assessment

In most watersheds there is likely to be a significant number of on-site treatment systems. At any time a percentage of these systems is likely to be failing. Effluent from failing systems will constitute a threat to public health and the environment if a sufficient volume of polluted water were to reach the waterway or groundwater. The greatest risk occurs during storm events, when heavy rain flushes polluted waters into streams; while, at the same time, the water treatment system is challenged by poor raw water quality. A recent study by the Cooperative Research Centre for Water Quality and Treatment (2007) reported that pollution indicators from poorly maintained systems were 100 to 1000 times higher than the best systems and that urban/intensive agriculture appeared to degrade water quality by factors of up to 1,000.

Hrudey & Hrudey (2004) documented cases of consumers being infected through the water supply system owing to the treatment processes being overtaxed, or human error in a storm event. For this reason most national and international guidelines recommend a multiple barrier approach to protect drinking water quality.

For watersheds with a significant number of on-site treatment systems, additional barriers may include built-in safety factors such as additional set-backs or buffer strips from waterways or wetlands downstream of concentrations of dwellings. This could be supplemented by a monitoring and compliance program to minimise the number of systems in failure mode at any time.

Watershed managers should hold a map locating all on-site systems within the watershed and identify and seek appropriate management controls for those likely to create the greatest risk because of their proximity to waterways or aquifer intake areas, surface geology, small lot size or localised high density.

New landowners are often unaware of the type of system serving their property. Maintenance is therefore likely to be neglected, especially if there is a significant recurring cost. New owners may also be unaware of the consequences of disposing of certain materials and chemicals into small on-site systems. In many instances, salespersons exaggerate the benefits, while down-playing the obligations of owning and operating these systems. This can lead to an unrealistic expectation as to the systems' performance and operating requirements.

Many on-site domestic plants are small activated sludge type plants relying on organisms in recirculated biosolids to remove organic material. These plants are often poor at treating peak loads and work best under constant load. Such plants are unsuited to holiday cottages or sites where peak loading can occur. At weekends these plants can become odorous and fail to treat the wastewater to the required specification until a sufficient concentration of biosolids has built up. This usually takes a few days, by which time the extra guests have departed.

The volume of wastewater to be disposed of is independent of the type of treatment system. Consequently the design of the disposal area and the hydraulic load will depend on the number of occupants and water consumption. In many instances a simple, low cost, septic system that has no moving parts and tolerates peak loads well may be appropriate if the disposal area is suitable.

The installation of low-flow water fittings and appliances within a property may reduce the volume of effluent to be managed but it is difficult to monitor properties to prevent new owners replacing these with normal equipment. In general, water consumption is more often a function of personal habits and culture than engineered devices. The frequency and duration of bathing and laundry, which are major water consuming activities, is not easily regulated. If systems fail, immediate action should be taken. The failure will often appear as clogging of soil in the effluent disposal area, by solids and/or grease, resulting in the surfacing and of site discharge of effluent. The cause, however, may be a failure of the treatment system or surcharge of fats and grease in a septic tank. It is prudent to have a reserve disposal field area that can be quickly established once the actual failure has been rectified. The approval process should require that this reserve area be protected from development.

Most on-site treatment regulations require adequate set-back from existing buildings and driveways. There is a risk, however, that, once established, future building extensions will encroach on the regulated set-back or even be located over the effluent field. The effluent field and surrounds should be identified and protected by some form of planning mechanism.

At any time a number of people living in or visiting the watershed will be infected by a contagious disease. This may include recently arrived overseas travellers. Because infecting organisms are usually present in bodily wastes it should always be assumed that effluent from on-site systems is highly infectious and appropriate systems put in place to manage this material.

For example, fodder irrigated with treated human sewage cannot be fed to pigs in Australia. This is a risk management procedure to prevent the spread of parasites that share pigs and humans as hosts. The parasites produce a spore that can be highly resistant and long-lived. Although these parasites are not present in Australia, the health authorities are concerned that tourists from infected countries could introduced these organisms into the local sewerage systems.

Watershed managers should take a similar risk management approach.

BEST PRACTICE

Design, Surveillance and Compliance

1. Any application for a multi-lot development that will not be connected to a sewerage system should include information that demonstrates to the satisfaction of the relevant authority that the site will be suitable for on-site disposal. This should include an assessment of the expected density of disposal systems and the limitations of the site. This work may, for example, set the minimum size of allotments in different parts of a proposed subdivision, based on geology, soil type and location of water courses.
2. Each proposed development should be required to provide a land capability assessment that identifies the constraints of the individual property. This should include, but not be limited to, an assessment of the lot size, soil type, its thickness, hydraulic capacity, depth to the water table, climate, distance from any waterway, location of any dam or bore, slope and likely maximum daily discharge volumes; as well as constraints caused by the proposed development. The assessor should determine a management plan for the site that will form the bases of enforceable use conditions if approval to proceed is granted.

3. A formal inspection of all new systems should be made before commissioning to ensure they meet the appropriate regulations and the design intent of the capability assessment.
4. The watershed manager or the statutory agency responsible for on-site systems should hold an up-to-date plan of all on-site systems under their jurisdiction and the approval document. This should be used to prioritise surveillance and compliance programs to target high risk systems.
5. The compliance program should include the facility to record and investigate complaints from other residents and the general community. Relying solely on regular inspections may result in failed systems continuing to discharge off site for months before they are detected.
6. Landowners should be required to forward copies of routine maintenance reports, effluent quality tests and maintenance activities (such as pumping out of septic systems). These can be used firstly to identify properties where maintenance is not being performed and secondly to follow up poorly operating systems.

Community Education

Landowners may not be aware of the importance of properly maintaining their on-site treatment system. New owners can be provided with brochures explaining the need for proper maintenance to protect their family's health as well as the health of the community. An effective mechanism is a routine inspection where the opportunity can be taken to reinforce the need to maintain the system and following the manufacturer's operating requirements to minimise health risks to the owner's family and the community and to avoid costly system failures.

Suitable printed material for landowners explaining their obligations, health risks to their family and likely costs of a failed system are useful.

Multiple Barriers

Because of the number of systems in a watershed and the reliance on individual owners for operation and maintenance, it is likely that a number of systems will be in failure mode at any point in time. This increases the risk of pollution of waterways in times of heavy rain. To allow for this, additional set-backs from waterways, longer than those normally required. constructed wetlands, and other measures should be considered to provide more barriers.

Density of Development

To prevent pollution of groundwater that may lead eventually to pollution of waterways or aquifer systems, an assessment that sets a safe, conservative density of on-site treatment systems should be undertaken. This may involve modelling or nutrient export calculations. Recent work on evaluating pathogen export for risk management models is becoming available (Ferguson 2006).

Monitoring

Watershed agencies should have a monitoring program in place that detects failures from areas of dense development. Further details of monitoring practices are provided in Chapter 7.

CASE STUDIES

Central Highlands Water (Australia)

Central Highlands Water, in conjunction with local municipalities, initiated a study of the condition of septic tanks within the local watershed. The study identified that a large percentage of septic tanks had thick fat and scum layers. Clearly many owners had not had their tanks pumped out for many years. Their belief was that because there was no odour or discharging, their systems did not need attention. The inspecting officer was able to explain that once that scum layer became too thick it would flow over the weir and block their infiltration lines. This would require a significant cost to repair. Most owners have now arranged for regular a pump-out.

Record Books in Victoria (Australia)

Some municipal councils in Victoria provide a record book with duplicate carbon copy pages to maintenance contractors. The contractor records details of systems pumped out or maintained and sends one copy to the municipality. This provides a mechanism to identify properties where maintenance is not being performed and triggers a follow-up. The contractors support the system because it encourages repeat business for them.

WHAT SHOULD WE KNOW?

1. To provide scientific support for controls on the density of development we need:
   * Better modelling tools to assess and predict the impact of increased development on water quality. This should include not only the impact of wastewater disposal but other sources such as storm runoff from urban and semi-urban areas, dust and refuse.
   * An economic model that recognises the community cost of poor watershed management in terms of increased treatment costs and health impacts.
2. A model bylaw or similar control to enforce mandatory auditing of on-site wastewater treatment systems in watersheds. Excellent mechanical aerated treatment systems exist but in most jurisdictions the operation and maintenance is left to the property owner. If the manufacturer's requirements for servicing and compliance checks on the effluent are not enforced these systems over time will fail to produce the expected effluent quality. While the life of the average dwelling may be in excess of 100 years mechanical wastewater treatment systems are likely to have an effective life of 25–35 years. Consequently a model system for audit and compliance is required.
3. Data of the fate of pathogens in local watersheds. Models are available but adequate data for their calibration is lacking.

OTHER RESOURCES

The National Onsite Wastewater Recycling Association, NOWRA, www.nowra.org

The Groundwater Foundation, www.groundwater.org/gi/wastewater.html

EPA Victoria, Australia, www.epa.vic.gov.au/water/wastewater/onsite.asp

REFERENCES

Cooperative Research Centre for Water Quality and Treatment. 2007. *Source Water Quality Assessment and the Management of Pathogens in Surface Water Catchments,* Research Report No. 29.

Ferguson C. 2006. *Deterministic Model of Microbial Sources, Fate and Transport: A Quantitative Tool for Pathogenic Catchment Budgeting,* Doctoral thesis, Dept of Biotechnology and Biomolecular Science, Faculty of Science, University of New South Wales.

Hrudey SE and Hrudey EJ. 2004. *Safe Drinking Water, Lessons from Recent Outbreaks in Affluent Nations,* IWA Publishing, London.

10 TOOLS AND RESOURCES

Pat Feehan

CONTEXT

Watershed managers need a variety of tools to assist them. The aim of this chapter is to provide a quick overview of some of the classes and types of tools available, and then to highlight some examples of useful tools.

Tools are required for each step of the management cycle, which includes assessment, planning, implementation, monitoring and evaluation. Modern organisations are called upon to be smart, fast, agile and responsive, and that is not possible without using good tools. Individuals working in watershed management know that it is becoming increasingly complex. It is unlikely that any one individual or organisation can command all the resources and competencies to be able to manage every aspect of a watershed. An appreciation of the scope of available resources and how to derive benefit from them is thus important.

OPTIONS

When practitioners think of tools they perhaps automatically think of computer-based tools such as models, GIS, visualisation tools, remotely sensed imagery and electronic databases. However, tools for watershed management can include a wide range of things such as management systems and frameworks (e.g., the Australian Drinking Water Guidelines), best management practices, communication tools (such as conceptual models) as well as tools for addressing the social and economic aspects of watershed management. The topic can include tools, systems or methods for turning data into information and knowledge; as well as achieving implementation of watershed management actions, plus data capture, warehousing and visualisation tools, risk management systems, knowledge management systems and electronic databases.

Watershed Models

Models are useful tool for watershed management: they help codify knowledge; can help managers determine the affect of particular management interventions and demonstrate to landholders where action will have greatest effect. However, when developing and applying models it's worth remembering:

- For every complex problem there is a solution that is simple, neat and wrong (attributed to H L Mencken)
- Technology is not always the best solution.

BEST PRACTICE

Catchment [Watershed] Modelling Toolkit

www.toolkit.net.au/

Modern watershed management requires that policy development, planning and intervention be undertaken in an integrated fashion, with consideration given to physical, ecological, economic and social systems. With this in mind, the Cooperative Research Centre for Catchment Hydrology (CRCCH) initiated the development of a 'catchment [watershed] modelling toolkit.' The ambition is to equip land and water managers, researchers and educators with an integrated collection of software tools and components designed to simulate holistic watershed responses to management and climate variability, at a range of scales and using a variety of approaches.

The Toolkit is a repository of software and supporting documentation intended to improve the efficiency and standard of watershed modelling. It is a resource for the land and water management industry and the research and development community.

The purpose of the Toolkit is to enable prediction of the multiple impacts of land and water management decisions at a whole-of-watershed scale. Its use will enable land and water managers to manage watersheds better, leading to improved environmental, economic and social outcomes for Australia.

The development of the toolkit concept was driven by the results of a survey (Marston et al. 1999) of Australian watershed managers, model users and model developers, to gauge the state of watershed modelling in Australia. Their survey revealed a range of problems that were limiting the efficiency and quality of watershed modelling. For instance, it was found that

- There was significant duplication of effort in model building
- The standard of computer code, interfaces and documentation underpinning watershed models was generally poor
- There were no agreed standards on how to code, document, quality assure and deliver watershed models to end-users
- Access to many watershed models was restricted
- There was virtually no holistic modelling being conducted at large spatial scales and no suitable software engineering paradigm for linking models.

The Toolkit has gradually been populated with a numberof models in the categories of

- Watershed Models
- Data Sets
- Data Analysis Tools
- Model Development
- River Tools
- Terrain Analysis Tools
- Urban Models
- Vegetation Models
- Water Quality Models
- Water Quantity Models
- Water Trading Models

Selecting Models

The CRCCH produced two papers in a series to assist a better understanding of watershed modelling and model selection. The papers discuss the major considerations in selecting an appropriate model for a particular application. Paper 1 is titled 'General approaches to modelling and practical issues of model choice' while the second looks at 'Water quality models — sediment and nutrients.' Both publications can be downloaded from the Catchment [Watershed] Modelling Toolkit web site at: www.toolkit.net.au/modelchoice.

Conceptual Diagrams

ian.umces.edu

Conceptual diagrams are effective tools to communicate complex messages in a simple and informative manner. In most cases they can be thought of as 'Thought (or Mind) Drawings.' Conceptual diagrams depict essential attributes of the system. The diagrams can evolve to capture an increasing understanding of the system. They essentially depict processes at different scales. An effective conceptual diagram has two key elements:

- Consistent symbols
- A comprehensive legend.

A good conceptual diagram will stand alone and be self-contained with no need to read accompanying explanation. Conceptual diagrams help clarify thinking and avoid ambiguity, and provide a communication interface by

- combining current scientific understanding with community priorities and environmental values;
- providing an approachable representation of the system; and
- allowing a shared vision to be created.

Conceptual diagrams are useful for integration and application by helping identify gaps, establish priorities and solicit an agreed synthesis. They are also easy to produce with click-and-drag symbol libraries in Adobe® Illustrator®, which enhances rapid iteration and ensures accurate depictions of system and processes. Applications for conceptual diagrams can include:

- Setting research agendas
- Developing scientific syntheses
- Setting monitoring programs
- Identifying management priorities
- Appearing in scientific literature; including peer-reviewed journal articles.

Integration and Application Network (IAN); The University of Maryland Center for Environmental Science (UMCES)

(www.umces.edu) Integration and Application Network (ian.umces.edu/) focuses on science integration and application. The IAN sites provides information about science communication and communication tools. An example is the IAN Symbol Library that contains over 1000 custom made symbols (in 28 categories) designed specifically for enhancing science communication skills for the graphically challenged. Using the click-and-drag capabilities of Adobe Illustrator, diagrammatic representations of complex processes can be developed easily with minimal graphical skills. The aim is to make them a standard resource for scientists, resource managers, community groups and environmentalists worldwide.

Arc Hydro — GIS for Water Resources

New tools are making it possible to address the complexities associated with these water resource challenges head on. The latest technological advancements are being based on Arc Hydro, the industry-adopted data model for water resources. This data model is providing a solid framework and acts as an "enabling" technology for efficient and collaborative water resource data analysis. The data model, the result of a collaboration between the Environmental Systems Research Institute (ESRI) and Dr. David Maidment of the Center for Research in Water Resources (CRWR) at the University of Texas, gives users a standardized template, which can be used for something as basic as hydrologic data storage, or for something as complex as executing regional interoperability and data sharing between local and state agencies with joint control over water resources.

The Arc Hydro data model can be defined as a geographic database containing a geologic information system (GIS) representation of a hydrological information system under a case-specific database design which is extensible, flexible, and adaptable to the user's requirements. It takes advantage of the next generation of spatial data in Relational Database Management Systems (an RDBMS-based GIS System), the geodatabase model. Conceptually, it is a combination of GIS objects enhanced with the capabilities of a relational database to allow for relationships, topologies, and geometric networks.

The ArcGIS Hydro data model describes geospatial and temporal data on surface water resource features of the landscape. The data model addresses three issues:

- Hydro Description — What are the principal water resource features of the landscape?
- Hydro Connectivity — How does water move from feature to feature?
- Hydro Modelling — What are the time patterns of water flow and water quality associated with these features?

For more information about Arc Hydro, go to www.crwr.utexas.edu/giswr/hydro/data/DataModel/archydromodel.pdf.

OTHER RESOURCES

Storage and aquatic ecosystem models

The University of Western Australia's Centre for Water Research has developed a series of models capable of describing the interaction between individual transport and mixing processes in stratified lakes, estuaries and coastal seas.

These fluid dynamic models can be coupled with Aquatic Ecosystem Dynamics models to allow for investigations into nutrient and metal cycling over a range of spatial and temporal scales across and within ecosystems: www.cwr.uwa.edu.au/.

Surf Your Watershed

The USEPA's Surf Your Watershed site is a service to help users locate, use, and share environmental information about a watershed: www.epa.gov/surf/.

Australian States provide similar information. For example, see the Watershed Information Mapper at nremap-sc.nre.vic.gov.au/MapShare.v2/imf.jsp?site=cim.

REFERENCE

Marston F, Argent R, Vertessy R, Cuddy S, and Rahman J. 2002. *The Status of Catchment Modelling in Australia,* CRC for Catchment Hydrology, Technical Report 02/4.

11 TOOLS FOR GROUNDWATER PROTECTION

Ali El-Kady

CONTEXT

This chapter is about groundwater sources: it explains source water assessment and protection and identifies available software that is suitable for analysis of groundwater flow and solute transport. A range of assumptions can be made to facilitate modelling and decision making, but the constraint of available data must be borne in mind, since sophisticated models are of little use if key parameters are unknown.

OPTIONS

Source Water Assessment and Protection

Assessing a water system's vulnerability to contamination is an important step toward the design and implementation of protection programs. Source water assessment and protection (SWAP) is a formal process to conduct an assessment and it consists of four steps:

- Delineating the source water assessment area, which can include either a fixed radius around the source or a capture zone that is estimated by computer models
- Preparing an inventory of potential contaminant sources within the assessment area
- Determining the public water system's susceptibility to contamination
- Ensuring public participation and outreach.

Groundwater Modelling

Groundwater models are useful tools for understanding various transport and fate processes, analyzing field data, designing the most efficient data collection scheme and predicting water levels and the migration pathways and concentrations of contaminants in groundwater. Thus such models can play a pivotal role in assessing source water vulnerability and in designing protection strategies. For example, models can be used in delineating protection zones described above in SWAP development. They can also be used in site-specific assessment of the chance of water contamination due to actual or hypothetical contaminant sources. Further, the models can be used in designing efficient cleanup and protection schemes.

However, the accuracy of model predictions depends upon many factors, including applicability of the groundwater flow and solute transport equations to the problem being simulated and successful model calibration and validation using available field data. In other words, model simulations depend on the conceptual model developed for the site, on the equations solved by the model software, and on the values of hydrogeological or geochemical parameters used in the model. Calibration entails defining sets of model parameters that provide model results that compare favorably with field measured hydraulic head or chemical concentrations. The lack of suitable data for certain parameters and of ways to estimate some others put severe limitations on model calibration. In such cases the calibration process is reduced to a fitting exercise, with questions regarding the model uniqueness, since various parameter combinations can provide similar fit between model results and field measurements.

Model validation would ideally include the successful matching of field results without any parameter adjustment. However, limitations regarding availability of independently measured parameters make it impossible to complete a successful model validation. Practical approaches for partial validation can include the use of available field data for model calibrations, then using the estimated parameters and the rest of the records in model validation. Sensitivity and uncertainty analyses are necessary elements of model predictions that aim at quantifying potential errors due to the lack of definite parameter values. Within these analyses, in many cases, final model input values are selected which result in a worst-case simulation, to be on the conservative side.

Approaches for Delineating Source Water Assessment Area

Delineated areas for source protection can be represented by the zone of influence, which is defined as the area from which the well draws water during pumping. This is dependent on the pumping rate and aquifer properties. They also can embody the zone of capture, which is defined as the up-gradient reaches of the aquifer from which the well receives water, and the size of which is dependent on the travel time. The areas can be defined by a remedial action zone, with a size designed to protect wells from unexpected contaminant release. Other possibilities include the use of an attenuation zone to provide an area to bring the concentrations of specific contaminants to desired levels by the time they reach the well head. Finally, the delineated area can consist of the well field or aquifer protection zone, which is defined as the area that includes all or part of the recharge zone for a particular well field or aquifer system.

The criteria for delineated areas design include distance away from the well, drawdown caused by pumping, travel time of potential contaminants reaching the well, the flow system or aquifer boundaries, or land use in the area. Design of delineated areas can include simple geometric shapes of a fixed arbitrary radius circumscribed around the water supply, based on a distance criterion. Such a method is not scientifically based and is generally used only in the early phases of a protection scheme or in combination with other methods. An improvement can be achieved by the use of an analytical equation to calculate the radius of a circular protection zone based on a time of travel criterion. Though still simple, this method provides better accuracy, depending on site conditions.

The next level uses simplified shapes that are based on analytical groundwater models and generally combines the flow boundary and time of travel criteria. Applying analytical models necessitates the use of restrictive assumptions that fail to recognize site-specific information, including aquifer variability and certain physical and chemical and transport processes.

At the top level, numerical models use mathematical approximations of groundwater flow and contaminant transport equations that can take into account a variety of hydrogeology and contamination conditions. These models offer possibly the most accurate delineations, though at a considerable cost in terms of time and resources. In addition, and as explained earlier, such solutions can be inaccurate if data for calibration and validation are lacking.

BEST PRACTICE

Best practice in this context means matching the modeling complexity and effort to the situation, in terms of its overall complexity, risk profile (see Chapter 4) and data availability. There are several groundwater models suitable for delineating a SWAP area. Most of the models adopt the time of travel criterion. After defining the flow field, particles are introduced to the system at the pumping well and allowed to travel in the exact opposite direction of groundwater flow for the specified travel time. The capture zone is defined as the area encompassing the stream lines created by the particle travel. Analytical, semi-analytical, or numerical approaches can be used to describe the flow field.

REGULATION

United States

In the United States, the Environmental Protection Agency (USEPA) mandates that each state develop a source water assessment program (SWAP) for public water systems. For groundwater sources, the assessment areas comprise two zones of 300 metres (1000 ft) and 1.6 kilometres (1 mile) fixed radii. In California, a SWAP includes four zones for groundwater sources, and an optional fifth zone. The well-site control zone encompasses the 15-metre radius immediately surrounding the well. Based on simplified approaches, charts are provided for estimating the sizes of other zones for travel times of 2, 5, and 10 years.

The State of Hawaii's SWAP differs from other states in the USA in that the use of site specific groundwater models in assessing source water areas and using hydrogeological data to the extent of their availability. The method can allow future updates as more information becomes available. The linkage between the models and a geographic information system allows utilization of spatially distributed data and presentation of assessment maps and susceptibility data in a convenient manner.

Australia

There are no national regulations for groundwater management in Australia, but the Australian Drinking Water Guidelines 2004 include the Framework for Drinking Water Quality Management, which implicitly addresses source water protection; and modelling could be inferred as necessary.

OTHER RESOURCES

Software for modeling groundwater flow; solute transport and for data handling is listed below. The list is not complete but it is comprehensive enough to cover the development of plans for groundwater resources assessment and protection. Free public-domain models are mainly available from US government agencies. The web sites for information about the software are provided and information about model acquisition is provided therein. Commercial software has the main advantage of a superior user interface, but at added cost.

It should be noted that the lists provide general groundwater flow and transport models that can be used not only for SWAP, but also for other groundwater applications. Other uses can include assessing water supply development, evaluating cases of contamination, and designing remediation strategies. Both saturated and unsaturated flow conditions are covered. The models deal with dissolved phases as well as non-aqueous phase liquids. The list covers both analytical and numerical solutions. Software for data analysis, such as geostatistical packages, is also listed.

- Model and modeling information: www.mindspring.com/~rbwinston/source.htm
- Interactive models: www.cee.uiuc.edu/transport/index.html

Free Public Domain Groundwater Software

Center for Subsurface Modeling Support (CSMoS), USEPA

CSMoS distributes various public domain ground-water and vadose zone models. Models and manuals are available through various different options. The following information is available from the CSMoS web site: www.epa.gov/ada/csmos/model-descr.html.

The site includes information about obtaining the software.

CSMoS also maintains an on-line database of commercially available groundwater modeling software: www.epa.gov/ada/csmos/omid.html.

Information below is obtained from the web site, where more can be found.

USEPA Wellhead Analytic Element Model

WhAEM2000 is a public domain, groundwater flow model designed to facilitate capture zone delineation and protection area mapping in support of the states' wellhead protection programs and source water assessment planning for public water supplies in the United States. www.epa.gov/ceampubl/gwater/whaem/index.htm

US Geological Survey Groundwater Software

Public domain subsurface models are available from the US Geological Survey website, where brief descriptions of available models are posted: water.usgs.gov/software/ground_water.html.

Commercial Software

Many different software tools are available from commercial suppliers. For examples see Groundwater Modelling System (GMS): www.ems-i.com/GMS/GMS_Overview/gms_overview.html.

Scientific Software Group (SSG) provides for environmental and water-resources engineers. SSG provides environmental software, groundwater software, groundwater modelling software, surface water software, air pollution software, geotechnical software, borehole logging software and more: www.scisoftware.com/.

Waterloo Hydrogeologic, Inc. www.waterloohydrogeologic.com/software/software_main.htm)

Argus Open Numerical Environments www.argusint.com/

12 COMMUNICATION

Chris Davis

CONTEXT

For any modern water business, communication is a critical function that underpins the success of the other technical and administrative activities. In watershed management, this critical function is even more marked because of the wide range of people affected by the business and potentially affecting the watershed; thus impacting water quality.

Long ago, management was viewed as an hierarchical, structured activity, modelled on the military; so top-down communication, on a need-to-know basis, was all that was needed. For a watershed manager, customers didn't exist, as such, but anyone doing anything in the watershed area could be managed by regulations, which, in turn, could be enforced by marshalls; in some cases, armed and in uniform. Communication didn't need to extend much beyond notices of prohibition or exclusion. Other stakeholders were of scant interest and communication with them was not important.

The world has changed now, so management is now a collaborative, team-based exercise, involving sharing of information of all types and an open-door approach to staff. The cooperation of watershed actors is sought by carefully structured messages, delivered by a range of channels; and collaboration with a myriad of other players is a fact of life; necessitating excellent, two-way communication. Chapter 3 discusses this issue in detail.

In short, a modern watershed manager must be an effective and fulsome communicator to have any hope of delivering on the remit to create the multiple barriers which protect water quality, and to achieve that in an affordable way. Transparency rules.

OPTIONS

Basic Approach

A basic choice to be made by a watershed manager is between openness and secrecy (or, at least, a need-to-know system) in terms of how it deals with staff, customers and other stakeholders. In some respects, a closed view is attractive, since it requires less work and seems to invite less criticism and/or dissent. On the other hand, trust in the watershed manager is a key goal; seldom reached without a high degree of openness. Likewise, collaboration with other actors in a watershed, and full engagement of staff, are both more likely to be achieved in a situation where information is made immediately and freely available. Finally, two-way communication is important to ensure that information vital to the watershed's operation is passed on to the manager, in a climate of mutual openness.

Stakeholder Communications

There is a range of stakeholders for watershed managers to communicate with, and each needs to be considered on its own merits, although some channels and messages will be in common.

Employees

Staff members are the first line of defence for the watershed: they must know what's expected of them, what's happening, and they must have excellent access to all necessary records which will throw light on their jobs. A shared sense of purpose with management is important, as is the ability to report and or discuss any issues which arise in their work.

Apart from technology and systems to ensure good records, good access and a feeling of accessibility, the critical ingredient is organisational culture, *the way we do things around here*. That culture comes from top management, which creates and then cements it by an open door approach to management and a timeous flow of information to staff. They are both critical to achieving the necessary level of communication.

Land Users

For the watershed manager, supportive and constructive behaviours by land users are crucial to success. Apart from the coercive power of regulations, ordinances and the like, a positive relationship and a shared set of values go a long way towards engendering the right attitudes. To achieve the desired climate, the watershed manager must ensure that goals and values are clearly enunciated and that there is a two-way communication with land users.

Although the Internet is today's key communication channel, it doesn't have universal reach, so other routes must be used to ensure that land users are both kept informed and are able to ask questions and to discuss issues with someone who will pay attention. The tools to achieve this will be a combination of the watershed manager's web site; newsletters; town hall meetings; a shop front with accessible customer service staff; and top management willing to use radio and television to get important points across when issues arise which are urgent and/or contentious. Depending on scale, one or another medium may be most apt. For rural areas, local radio talkback shows often have strong penetration and can be used to conduct meaningful dialogues at little cost.

Staff used to carry out monitoring (Chapter 7) and surveillance (Chapter 6) need to wear corporate clothing and drive branded vehicles, so land users get to know them and are willing to ask questions or to share information.

Customers

Watershed managers vary institutionally in that some have retail customers, while others sell bulk water to distributors, so may have just one or a handful of retailers as customers. In principle, though, the individual water consumers are still indirectly linked to the watershed and need to be treated as customers, despite the fact that the formal connection might be via a third party. Customers need to be kept well informed about watershed issues: quality status; supply certainty; risk management and, in extremis, emergencies.

The channels to reach customers include bill stuffers (leaflets inserted with water accounts); the Internet; newsletters (print or electronic); shopfronts; call centres; print press; radio; television and customer liaison committees. A watershed manager needs to strive for a strong, trusted brand, and must tailor all communications to be consistent with the brand, its values and customers' cultural preferences.

In terms of mechanics and details, this is essentially a marketing exercise, but it cannot succeed without real integrity and good faith on the part of the watershed manager. Customers may not be watershed experts, but they are not stupid, so dishonesty will be detected and will have negative results.

Other Agencies

It is critical for a watershed manager to be in close contact with, and to win the constructive collaboration of, all other agencies with a stake in the watershed. This includes planning (See Chapter 5), other utilities, local authorities (where the manager is not also the local authority), the environment protection agency, agriculture department, police, national parks, transport (especially roads), natural resource management and relevant NGOs.

The precise nature of interaction and communication will be dictated by the importance of the parties' successful collaboration. Critical agencies need good relations between CEOs, while low priority agencies may just need good field-level contacts.

Industry and Professionals

Watershed managers should also be actively connected to water industry peers and to practitioners. The links maintained can be invaluable in terms of keeping industry actors in touch with what the watershed manager is doing, as well as connecting the watershed manager's staff to the community of practice.

In the United States, Canada, Australia and New Zealand there are both industry associations and professional interest associations which offer value in the communication field. These include AWWA, AWA, NZWWA, IWA, WSAA, AMWA, and others. All are easily web accessible.

BEST PRACTICE

Planned and Structured Approach

Best practice is a combination of effective communication tools, used in a coherent way to ensure that staff, land users, customers and other stakeholders are all well informed about the activities, needs, plans, policies, values and operating results of the watershed manager. The communication function is important enough to warrant a second-tier manager (for all except the smallest watershed management operations) and a corporate plan for its successful execution.

Peer Comparisons

Ensuring value for money is important. On the other hand, parsimony in terms of outlays, and short-changing communications efforts, are likely to lead to problems with perceptions and a lack of understanding among the intended audiences. Comparing notes with peers and benchmarking are two ways to calibrate an agency's communications efforts against those of others.

In Australia, for example, major utilities' communications managers meet regularly to discuss their common interests, in a sub-group of the Water Services Association. Also in Australia, the Water Education Network (WEN) is operated by the Australian Water Association; to bring together community education practitioners and enable them to share ideas and learnings. Apart from a web page (www.awa.asn.au/wen/), WEN holds conferences on a biennial basis, to foster the exchange of ideas.

Surveys

A scientific approach to tracking communications success is necessary:

- For employees: independent, confidential (i.e., anonymous) surveys are needed to monitor morale in regard to communication and acculturation
- For customers: longitudinal surveys, focus groups and the logging of complaints and queries; a key determinant is the level of trust among customers
- For other stakeholders: it is important to demonstrate that communication has been effective in avoiding clashes of direction or glitches arising from lapses in information dissemination.

Consistency of survey contents over time is important so that longitudinal changes can be tracked. Credibility is important, so employing professionally trained staff, or engaging reputable survey firms is necessary.

Peer Connections

Being actively engaged in relevant professional and industry groups is symptomatic of a healthy, open and transparent watershed manager. This also applies in the area of research, which generally embraces a communication dimension.

CASE STUDIES

Sydney Catchment Authority

The Sydney Catchment Authority (SCA) was established by the New South Wales government after the notorious *Cryptosporidium* scare in 1998 as a single-purpose watershed manager for all the surface water sources serving the 4.2 million population of Sydney, the Blue Mountains and the Illawarra in New South Wales, Australia. As a bulk water supplier to Sydney Water, the SCA has to reach its ultimate customers through its own channels and via Sydney Water. A central plank to its communication is its web site (www.sca.nsw.gov.au) and that site is well linked to Sydney Water's site (www.sydneywater.com.au).

The SCA works closely with Sydney Water on the consumer communications and in terms of liaising with other agencies, especially health. The SCA is managed under the Environment portfolio of the NSW Government, so is well connected there, but works closely with Sydney Water and the NSW Health Department on the health dimension. For land users, however, SCA makes use of its own staff and resources to maintain effective links.

In terms of industry communications, the SCA is an active sponsor of industry events — conferences, seminars and the like — so it is well connected to current thought and practice.

WHAT SHOULD WE KNOW?

Although quite a few studies have been conducted into consumer perceptions about water, the specific communication needs of watershed managers have not been widely studied, so there is scope for research and information sharing in that area. The collaborative efforts which were the genesis of this book could, perhaps, also lead to future information sharing among members of the AWA and AWWA, without the formal need to initiate research as such.

OTHER RESOURCES

The AWWA Bookstore (www.awwa.org/bookstore) carries the following titles which are relevant:

- *Strategic Communication Planning: A Guide for Water Utilities Customer Attitudes, Behavior, and the Impact of Communications Efforts*
- *Message Management: Effective Communications*
- *Customer Attitudes, Behavior, and the Impact of Communication Efforts*
- M5, *Water Utility Management*
- *Focus First on Service*

The Institute for Sustainable Futures (ISF) at the University of Technology in Sydney (UTS) conducts social research into all dimensions of water system operation, including watershed management. www.isf.uts.edu.au

The Australian Research Centre for Water in Society, ARCWIS, researches community views and issues like communication: www.clw.csiro.au/research/society/arcwis/. ARCWIS is operated by the national science agency, the Commonwealth Scientific and Industrial Research Organisation (CSIRO).

13 INCIDENT MANAGEMENT

Chris Davis

CONTEXT

Every watershed is vulnerable to some degree to unexpected events which could compromise water quality; ranging from bushfires to terrorist action. The likelihood ranges from remote (say an aircraft crashing) to high (vandalism, which seems to be endemic) and the potential harm ranges from minor (a short interruption to supply) to catastrophic (complete loss of supply for an extended period, or serious health impacts on consumers). In terms of risk, each potential threat needs to be ranked in terms of the product of the likelihood and the hazard. These concepts of risk characterisation and assessment are addressed in Chapter 4, so this chapter is devoted to incident management as such.

By its nature, incident management involves preparing for events which may never happen, but which could be critical to the water quality and the health of consumers served by a watershed. This requires meticulous planning and preparation and, as the event may never actually occur, realistic rehearsals to ensure that all the people and systems involved react appropriately when something untoward does happen.

All stakeholders involved with a watershed must know how they fit into the picture when an incident occurs. This includes all utility workers (i.e., power, gas and communications, as well as water), emergency response teams, police, health workers, and the army. There has to be an explicit chain of command so, when a given incident happens, there is no ambiguity about who reports to whom and, most important, who is in overall charge.

Incidents are inherently unpredictable, so contingencies have to be allowed for with imagination and foresight. A key issue, for example, is the safety of workers' families. There are examples of firefighters (often volunteers) whose homes have been burned down while they are fighting fires themselves. Ensuring that workers' loved ones are safe is a fundamental plank of an effective emergency system.

The risk management and mitigation process is iterative and is continuous. Identified risks have to be 'treated.' The treatment might mean obviating the risk (e.g., padlocking entries to reservoirs) or planning to deal with it when it happens (e.g., cleaning up and containing a chemical spill). The quantitative risk assessment process will help to guide watershed managers on how much should be invested in avoidance and how much in response.

OPTIONS

Much of what incident managers must do is not optional, since it is straightforward and essential, but there are options in terms of the extent to which desk-top and field exercises are carried out to test systems and responses.

The central components of an incident response system are protective measures and a carefully worked out response plan which is current and known to everyone involved.

Protective Measures

Based on a thorough knowledge of the watershed, a manager must ensure that all reasonable steps to avoid incidents have been taken. This is a balancing act between doing nothing but incurring some significant risks, on one hand, and investing so much in protective steps that it is excessive, on the other. Probably the most common incident threats to water quality in a watershed will be:

- Vandalism and trespassing
- Accidental spillages from vehicle accidents
- Fires
- Floods.

Each of these, and all the less common incidents, must be considered in the light of what provisions might be made to prevent them or minimise their impact. Thus, fencing off and securing sensitive areas; re-aligning roads and crash barriers; constructing firebreaks and building flood protection devices must all be considered as options. Clearly, there is no merit in investing more on the measure than the cost of dealing with an actual incident, and many interventions that seem attractive may simply be beyond the means of the watershed manager.

In terms of the old Australian adage; it is 'horses for courses' — every measure has to be tailored to local circumstances. In terms of due diligence, it is important to document the process of deliberation leading up to a decision to make a provision or not to do so. A rigorous, quantitative risk assessment is the best evidence to demonstrate how decisions have been reached. As suggested in Chapter 4, however, often it is only feasible to conduct a semi-quantitative assessment.

The watershed poses particular challenges for incident management, since it is intrinsically a widely spread geographic area, home to a wide variety of residences and activities. The variety of incidents which might occur in a watershed is vast, so imagination is needed to consider possibilities, and agility and inventiveness are necessary to address incidents as they happen.

Response Plan

A response plan must be worked up by the watershed manager, in consultation with all relevant actors; and must be kept current and, of course, accessible to and known by all responsible officials. The plan must include the following:

- Lists of all key personnel in the watershed managing agency, as well as all the relevant contacts in other agencies: firefighting, emergency services, police, medical, laboratories, army, etc.
- Designated incident managers for each potential condition, with the definitions and designations well known by all concerned to avoid any ambiguity if and when an incident does happen
- An incident room, in a safe place and having all the necessary communications equipment, information (maps, contacts, etc.) and facilities to sustain the operation until the incident has been resolved (of course, minor incidents won't require a room)
- Communication protocols to inform all stakeholders, and especially the community, via the media. For major incidents, pre-prepared media releases may be necessary to cut down time lost between incident detection and effective public communication
- Physical equipment and stocks to deal with the incident. For example, a water quality incident may demand an alternative source of drinking water, so access to tankers and an agreement with a neighbouring utility need to be lined up
- For serious incidents, such as flood or fire, alternative plans for

essential services (transport, roads and communications) to ensure that necessary support is at hand

- The watershed manager must ensure that emergency responses by other services to their own circumstances include for involving the watershed management team to whatever extent is necessary
- Testing facilities need to be lined up in case any unusual parameters must be analysed for in water; sampling protocols must be worked out and appropriate responses planned in case something specially challenging is discovered
- Access to toxicology databases should be arranged, so no time is wasted searching for facts about toxic or biological agents
- Microtox or equivalent sampling equipment should be located so it can be deployed quickly and effectively if the need to assess toxicity arises
- In the case of any malevolent act, the need to involve either the police or military and, if necessary, counter-insurgency forces must be allowed for and necessary lines of communication established in advance.

Practice and Rehearsals

To be sure that the response to an incident does go according to plan, it is necessary to train staff in the utility as well as those in other affected organisations. Not only does practice prepare people, but it also reveals deficiencies in the response plans which may not be apparent without an actual rehearsal. As always, cost and ramifications have to be borne in mind; some incidents are simply beyond the scope of a utility, while others happen so often that practising is redundant.

When physical rehearsals are deemed necessary, it is important to think through who will know it's just an exercise and who will be reacting under the impression that a real emergency is under way. If drastic reactions and emotions may be elicited, caution should be exercised in allowing participants to believe in a real emergency. When a practice event is over, all involved should be debriefed effectively to extract the learnings and update the response plan accordingly.

BEST PRACTICE

Best practice in watershed incident management is a living and tested management cycle of Plan-Do-Act-Review, designed to address any foreseeable incident, but flexible enough to address the unforeseen.

A key item in the loop is a risk assessment (Chapter 4) which must be current and relevant.

Response plans must be formulated, tested and practised (either physically, or as a desk exercise) and known to all internal and external stakeholders.

Up-to-date contact lists for all actors involved must be easily accessible and, ideally, all key players should be personally known to one another.

The chain of command must be pre-arranged and explicitly understood, to avoid any gaps or overlaps in the event of a real incident. For a given incident type, the person in charge should be the most apt for that particular incident, not the most senior person available.

When an incident is over, there should be an agreed endpoint, at which normal management resumes; if not there is the risk of a dragged-out, ambiguous situation and ongoing confusion.

All incidents, practices, minor, major, real and hoaxes, must be addressed professionally. The AWWA manual, M19, *Emergency Planning for Water Utilities,* fourth edition (2001) includes a neat, iterative cycle for managing incidents that is summarised in the following list:

1. Analyse the type and severity of the emergency — Based on the previously conducted hazard summary and system vulnerability analysis, the results of this analysis should lead to the most prudent and effective response.
2. Provide emergency assistance to save

lives — Initiate immediate actions required to safeguard the lives and health of employees and the public; 'first evacuate, then repair.'

3. Reduce the probability of additional injuries or damage — When emergency conditions threatening immediate injury or loss of life have been eliminated, repairs or temporary mitigation actions should be initiated. During this phase, the public should be kept appraised of boil water orders or other emergency notifications.
4. Perform emergency repairs based on priority demand — The next step is to carry out repairs or other actions which will provide services to priority customers or meet priority demands [this is not specially relevant to watersheds].
5. Return system to normal levels (recovery) — Finally, as control is regained, perform less urgent repairs in order of priority. Lift boil-water or reduced-use orders only when the quality and quantity of the water supply is assured.
6. Evaluate response and preparedness plan — Even during a crisis, a utility should review its responses and compare them to those outlined in the preparedness plan. Preplanned actions may not be as effective as anticipated and real-time modifications may be in order. Once the crisis has passed and the system is operating normally, the preparedness plan, response actions, and outcomes should be thoroughly evaluated.
7. Revise plan as necessary — Following the 'lessons learned' evaluation, it may be necessary to modify the preparedness plan. Any changes to the plan, response actions, or facilities should be made promptly. Remember that the response plan should not reflect any facility or equipment changes until they occur.

CASE STUDY

In 2002, a distribution reservoir in the Northern Territory (Australia) town of Tennant Creek (population 3,000) was tampered with, leading to concerns of sabotage. Air ducts on the roof were broken off and a substance was spilled across the roof. Valves within the fenced tank compound were also tampered with — these in fact isolated the tank, and via the SCADA system, this isolation was the first indication that something was wrong.

The local police were notified, and immediately established a crime scene at the reservoir site. There are numerous precedents that disgruntled employees are among the most likely to sabotage existing infrastructure, hence the first reaction of the police was to detain and interview Power and Water staff. With only three permanent water services staff in Tennant Creek, this immediately restricted the organisation's ability to respond and manage the incident. Limited understanding of how both police and Power and Water would respond also led to confusion — each assumed that the other would be able to provide relatively rapid assessment and evidence of what contamination may have occurred.

In the end, the incident was found to be a case of vandalism, so no actual harm occurred. However, it illustrates how important it is to have all actors involved with a water system included in a full appreciation of what needs to be done, who is in charge and how to communicate when an incident happens.

OTHER RESOURCES

Information on incident management is not always made freely available, for obvious reasons; but, in both the United States and Australia there are secure networks among stakeholders, between whom information is freely shared.

Emergency Management Australia is a national agency set up to support everyone responsible for emergencies: www.ema.gov.au.

AWWA publications, available from the Bookstore (www.awwa.org/bookstore), include:

- M19, *Emergency Planning for Water Utilities*
- *Emergency Planning: the Big Picture for Water Utilities* (video).

WHAT SHOULD WE KNOW?

There is always room for more information about incidents and cost effective ways of dealing with them. Empirical data about how incidents were handled and with what success would help incident managers hone their plans and rehearsals, to maximise the likelihood of a successful outcome when a real-life incident occurs. Active participation in local emergency networks is probably the best way to gain such knowledge.

CONTRIBUTORS

Michael Anderson, University of California, Riverside, Michael.Anderson@ucr.edu

Chris Davis, University of Technology, Sydney, Chris.Davis@uts.edu.au

Daniel Deere, Water Futures & CRC Water Quality and Treatment, dandeere@bigpond.net.au

Ali El-Kady, University of Hawaii, elkadi@hawaii.edu

Pat Feehan, Feehan Consulting, pfeehan@mcmedia.com.au

Robert Ford, bob.ford@ncable.net.au

Rhonda Harris, PRO-OPS, Inc., rharris@pro-ops.net

Rod Holme, rholme@sympatico.ca

Clairly Lance, Water Corporation of WA, Clairly.Lance@watercorporation.com.au

Rachael Miller, Water Corporation of WA, Rachael.Miller@watercorporation.com.au

Jerry Schulte, ORSANCO, jschulte@orsanco.org

Ariamalar Selvakumar, US Environmental Protection Agency, Selvakumar.Ariamalar@epamail.epa.gov

Perri Standish-Lee, Black & Veatch, Standish-leepp@bv.com

Joyce Perdek Walling, US Environmental Protection Agency, Walling.Joyce@epa.gov

Rob Walton, Oxford County, Canada, rwalton@county.oxford.on.ca

Bruce Whitehill, Sydney Catchment Authority, Bruce.Whitehill@sca.nsw.gov.au

CONVERSION FACTORS

U.S. to Metric

1 mile = 1.609 km
1 foot = 0.305 m
1 inch = 25.4 mm
1 acre = 0.405 hectares
1 acre-foot = 1.233 megalitres
1 US gallon = 3.785 litres

Metric to U.S.

1 km = 0.621 miles
1 m = 3.281 ft
1 mm = 0.0394 inches
1 hectare = 2.471 acres
1 ML = 0.811 acre-feet
1 GL = 811 acre-feet
1 L = 0.264 U.S. gallons

For painless conversions, online, visit www.onlineconversion.com

INDEX

f = figure; *t* = table